Soil Landscape Analysis

SOIL LANDSCAPE ANALYSIS

FRANCIS D. HOLE

JAMES B. CAMPBELL

ROWMAN & ALLANHELD
PUBLISHERS

ROWMAN & ALLANHELD

Published in the United States of America in 1985
by Rowman & Allanheld, Publishers
(A division of Littlefield, Adams & Company)
81 Adams Drive, Totowa, New Jersey 07512

Library of Congress Cataloging in Publication Data

Hole, Francis Doan, 1913–
Soil landscape analysis

Bibliography: p.
Includes index.
1. Soil geography. 2. Soils. 3. Landscape.
I. Campbell, James B., 1944– . II. Title.
S591.H69 1984 641.4′9 83–24418
ISBN 0–86598–140-X

85 86 / 10 9 8 7 6 5 4 3 2 1

Printed in the United States of America

We dedicate this volume to

V.M. Fridland (1919–1983)
Soil Geographer

Contents

Tables

Illustrations

Figures

Plates

Preface

Characterization of soil cover patterns by soil landscape analysis is a developing science that has roots in the work of Dokuchaev, Sibirtsev, Hilgard, and other early pedologists. Impressive contributions have been made recently by V.M. Fridland and his co-workers to this important subdiscipline of soil science. At the same time that study of the *soil profile* (and *pedons*) has been emphasized, to the benefit of modern soil taxonomy, delineation of soil landscape patterns has been extended over a greater area each year. For example, publication of county soil surveys by the U.S. Soil Conservation Service, and the U.S. Forest Service, in cooperation with universities and other agencies has made available a vast amount of geographic information, especially in the last two decades. Numerical analysis may be used to characterize domains of the soil continuum at various scales. Re-examination of the soil landscape in the field gives us an opportunity to add ecological and statistical observations of the soil cover.

We have defined the subject of this book as the study of the soil landscape from a geographic perspective. Any subject can be examined from many disciplinary perspectives, and within disciplines, from varied national traditions and personal viewpoints. We have attempted to cross as many of these disciplinary, personal, and national boundaries as we could. We view this book as a statement of a perspective: not so much a series of facts as a style of examining the pedological landscape. Its contents form a series of propositions to be examined, evaluated, debated, and then perhaps modified in the light of subsequent experience.

This book can claim only to introduce a complex subject, on which some of our teaching has been focused in university courses of study. We hope that our thoughts and our efforts will be of interest to the many students of soil geography throughout the world, and will stimulate studies that will be more comprehensive than this one.

Finally, just as we completed preparation of the last draft of our manuscript, we received news of the death of V.M. Fridland, to whom we have dedicated this volume. Only a few days earlier we (FDH) had received what must have been one of his last letters (confirming biographical information for this book); our reply probably did not reach him. We are saddened by the loss of one of the most stimulating and productive scientists in the field of soil landscape studies. Perhaps this volume may help to consolidate some of his contributions.

Francis D. Hole
Madison, Wisconsin

James B. Campbell
Blacksburg, Virginia

1

A Perspective for Landscape Studies

The science of *pedology* (soil science) has been a late arrival among the natural sciences. It is said to have been founded by V.V. Dokuchaev (1846–1903) and his students in Russia and by E.W. Hilgard (1833–1916) in the United States (Jenny, 1961). A century-old science is a young science. The disciplines of mineralogy, botany, and zoology have been developing about three times as long. One reason for the late blooming of soil science is the transitional nature of the soil blanket over much of the land of this planet. This cover lies at the bottom of the atmosphere, between green vegetation and bedrock and other geologic materials. Furthermore, soil has escaped discriminating observation because it is everywhere underfoot and is commonplace for most of us. "It takes a genius to undertake the analysis of the obvious" wrote Whitehead (1925). The fact that most of the *soil cover* is actually hidden from sight makes this resource an unlikely subject for systematic inquiry. A simple comparison of words that we use with respect to trees, waters, and soils indicates the rudimentary nature of folk vocabulary about soils (Table 1.1). The common words about soils seem to refer to materials and conditions of soils, not to distinct individuals and groups of individuals. Folk terms for particular kinds of "lands" ("Chernozem" for example) are inclusive of other features besides the soils, particularly vegetation. Since soils are not visible as free-standing entities, we may be expected to be slow to recognize them as objects of study.

The founders of pedology observed that there are contrasting zones of soil that correspond generally to major phyto-climatic zones. These early observers examined and described sequences of *soil horizons,* the soil layers that were exposed in local cross-sections of the soil blanket. Soil zones and characteristic *soil profiles* (see Figure 1.1) of component soils were differentiated. It was a little like looking at plugs taken from so many melons of different kinds. Detailed observations were concentrated at representative points in landscapes; to dissect a whole soil landscape was not only physically impossible but also undesirable because disturbance deteriorates natural soil structure.

The concepts of the two-dimensional soil profile, the three-dimensional *soil pedon* and component *peds* (the natural separable units within soil hor-

Table 1.1 A Simple Comparison of Common Words Used With Respect to Soils and Two Other Natural Resources

Category	Comparative vocabularies		
	Trees	Waters	Soils
Material	Wood	Water	Soil, earth, ground, mud, muck, peat, sand, silt, clay, dirt
Distinct individual	Tree	Lake, river, sea, ocean, falls	–
Group of individuals	Forest, thicket, copse, grove, woods	Chain of lakes, network of streams	Black land (Chernozem) White earth (Podzol)

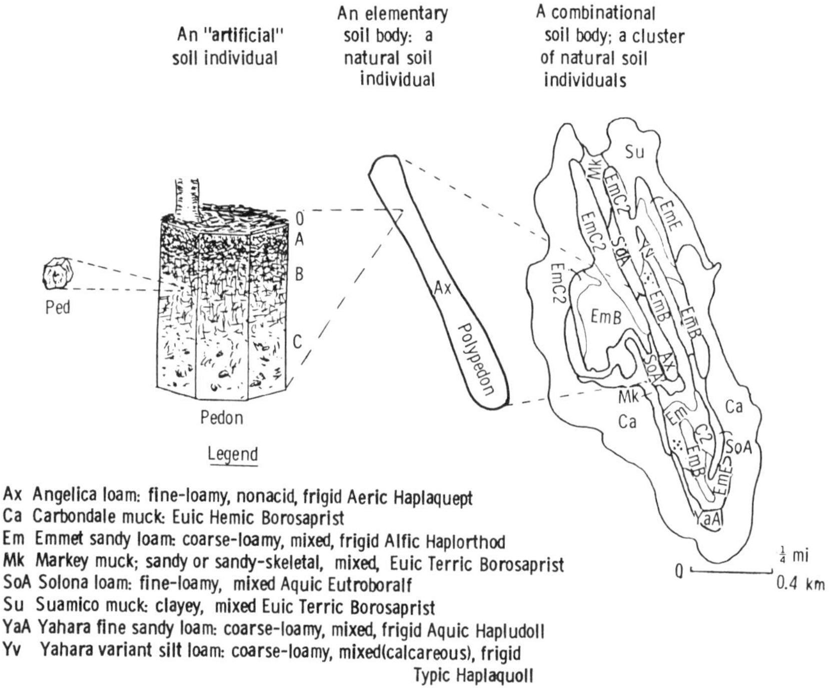

Figure 1.1. The concepts of ped, pedon, elementary soil body (polypeden), and combinational soil body. Any vertical face of the pedon can serve as a two-dimensional soil profile. The combinational soil body is taken from maps by Link et al. (1978). The relief is 12 meters (40 ft), between elevations of 200 and 212 meters (665–705 ft), on the Door Peninsula, Wisconsin, on the west shore of Lake Michigan.

izons) evolved in order to focus attention systematically on pedological units of study (Figure 1.1). A considerable leap of imagination was made when the concepts of an elementary soil body *(polypedon)* and a cluster of such bodies *(combinational soil body)* were formulated.

A *soil body,* the basic soil individual, is infinitesimally smaller than a soil zone. Even the vast bodies of *Chernozem* do not compare in size with a major soil zone. Soil body Ax in Figure 1.1 is about 0.4 km long and is a cartographic representation of a complete specimen of a soil. It is a particular example of an approximation of the ideal cluster of like pedons called the polypedon. One can compare a natural soil body to a tree by saying that each of these entities is a complete individual, and that each has a considerable part out of sight. Along boundaries the hidden parts interpenetrate in varying degrees with adjacent individuals. A soil body is bounded by other bodies of soil and perhaps also by *not-soil* (rock outcrops, lakes, seas, glaciers, salt flats). Comparing a soil body to a water body may be more appropriate, because it is possible for a water body, such as a lake, to evolve into a soil body by processes of slow accumulation of peat in it, and of subsequent "ripening." The resulting body of peat may have about the same shape and volume as the lake did orginally. If the lake were a large one, perhaps 10 km across, the ultimate body of peaty material would likely be a cluster of peat bodies differing in kind and arrangement of materials, in water regime and other conditions.

Equipped with at least a working definition of a soil body, soil mappers have for about a century been producing maps that depict with some degree of generalization the intricate pattern of soil bodies in the soil cover or soil mantle. In the process, soil scientists have given first priority to soil horizons in terms of their properties and use-related qualities. Less attention has been devoted to study of landscape properties of soils. For example, two kinds of Fayette silt loam (Fine-silty mixed mesic Typic Hapludalfs) [1] have been recognized in the Driftless Area of Wisconsin: Fayette silt loam on upland ridges and Fayette silt loam on colluvial footslopes or benches. Yet the soil profiles are different. The second soil has some stratification in the B and C horizons and contains some stones. The ridgetop soil lacks these features (Slota and Garvey, 1961). Soil maps lead the unspecialized reader to assume that each soil body, like a rather simple water body, is reasonably uniform and that the boundaries shown are at least as abrupt as the shores of a lake (Plate 1.1). This is usually not the case. [2] Therefore, a complete explanation of a soil landscape would include a statement about: (1) the degree to which bodies of each kind of soil, as depicted on a map, are impure as the result of inclusion of patches of other kinds of soil, and (2) the degree of definiteness of the boundaries between soil bodies. [3] This kind of information just is not widely available. Hence, most modern soil maps, however useful, are first approximations. The need for much more work in soil geography is evident. The principal purpose of this book is to present the subject of soil geography in terms of the analysis of soil landscapes and of maps of them, and to report some results of such analyses.

Soil geography is the last mentioned and probably the least developed of the subdisciplines of pedology (soil science) listed by Haase (1968): (1) soil

Plate 1.1 Contact between two soils, Green Lake County, Wisconsin. *Right*: Kidder fine sandy loam (Typic Hapludalf): *Left*: Ritchey fine sandy loam (Lithic Hapludalf). This extraordinarily sharp boundary is much more easily observed than are the gradual boundaries that are much more typical in nature. (Photo by J.B. Campbell.)

anatomy (morphology), (2) soil physiology (processes), (3) soil genesis, (4) soil systematics (taxonomy), and (5) *soil geography*. The carefully considered decision to omit from *Soil Taxonomy* (Soil Survey Staff, 1975a) *phases* (subdivisions of *soil series)* of soil erosion, slope, and surface conditions including texture (Buol, 1976) was in a sense a placement of these ecologically important features in the discipline of soil geography for investigation and explanation. Fridland (1976a, 1976b) and his co-workers have long been world leaders in the analysis of soil cover and the soil combinations in it for both scientific and practical purposes.

A POINT OF DEPARTURE

Just as an elementary soil body is a natural, relatively distinct cluster of pedons, so a combinational soil body is a natural distinct cluster of elementary soil bodies. A combinational soil body is a distinct unit in the soil cover. It is the pedological portion of a particular *terrain* that consists of a unique aggregation of soil bodies (Figure 1.2). *Landscape* has a pedologic structure (pattern), which means that it is characterized by fabric, i.e., size, shape, and arrangement of component soil bodies (soil inventory, or "SI," components according to Dan H. Yaalon, personal communication, 1978). Although the terms polypedon and soil body have been used interchangeably by some

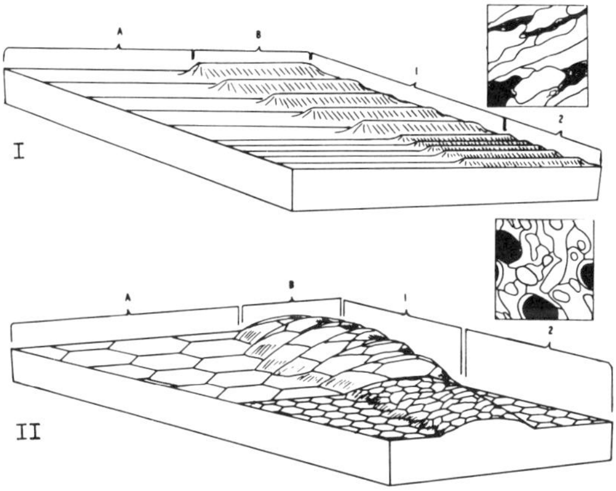

Figure 1.2. Diagrams showing eight idealized combinational soil bodies. Unit I: four bodies with parallel linear patterns, level (A), hilly (B), relatively broad (1), and narrow (2). Portion of soil map (*right*) shows a linear trend to soil bodies in Oklahoma. Unit II: four equidimensional elementary soil bodies, level (A), sloping (B), relatively large (1), and small (2). Map shows parts of three oval soil bodies set in a matrix of smaller oval and linear shapes in the Carolina Bays area of North Carolina. Shaded areas in 1 km sq maps indicate poor natural drainage.

workers, there is a distinct difference in flavor, and in meaning. The term polypedon is used here in a conceptual or idealistic sense. Soil body refers to an actual pedological entity in a landscape. These concepts and terms have been discussed by Buol, Hole, and McCracken (1980). However, their use of the term *soilscape* is not followed here in the interest of greater objectivity.

Soil geography is concerned with the nature and genesis of both elementary and combinational soil bodies, and of interactions between them and other environmental entities, including human beings (Figure 1.3).

Genesis of soil bodies is that phase of soil science that deals with factors and processes by which components and patterns of soil terrain form and change. It includes description and interpretation of cross-sections along soil transects, and of areal patterns of soils. Soil terrain genesis is the study of origin and change in these patterns. This involves a study of entire soil bodies in their complete thicknesses and lateral extent. The recent practice of publication on an aerial photographic base of numerous detailed soil maps of counties and special areas of the United States provides an opportunity for development and elucidation of soil geography. The fact that these maps are generalizations of the real pedological pattern on the land motivates the soil geographer to re-examine the soil terrain patterns in the field.

Guy D. Smith (1907–1981), chief architect of the new taxonomy (Soil Survey Staff, 1975a), stated that "*Soil Taxonomy* does not concern itself with

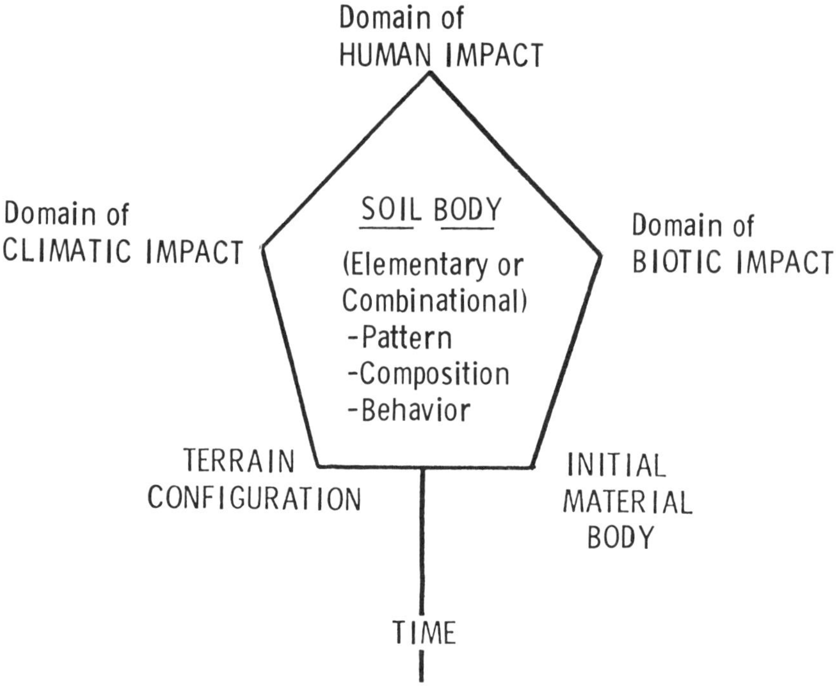

Figure 1.3. A pentagon showing geographic factors of soil body formation.

areal extent," and that "soil landscape properties are probably not well defined" (Smith and Leamy, 1981). Knox (1965, p.83) concluded that "future developments may lead to a soil classification system based upon some kind of landscape units, but at present the difficulties seem overwhelming." Hammond (1962, p. 72) called attention to the need for a component-characteristic organization of landscape analysis: "Landform description stands alone in its stubborn adherence to a genetic organization."

BACKGROUND OF SOILSCAPE DEVELOPMENT THEORY

Undoubtedly geomorphologists and pedologists have, since the days of Alexander von Humboldt (1769–1859) had images in their minds of soilscape (soil landscape) domains. The expression of such images in maps, graphs, words and numbers has been an evolutionary process fostered by the demands and opportunities of soil survey. In 1883 Professor T.C. Chamberlin wrote at the beginning of a chapter on soils in a treatise on the geology of Wisconsin, U.S.A.: "There are few subjects upon which it is more difficult to make an accurate, and at the same time intelligible report, than upon soils. The difficulty arises partly from the nature of the subject and partly from the vagueness of the terms used in speaking of soils" (Vol. II, p. 88). During the ensuing century, definition of terms concerning soil materials, soil horizons and soil species has advanced impressively in many countries.

The reasons that geographic aspects of the *pedosphere* (the thin film of soil on the land, forming a discontinuous hollow sphere on our planet) have been less codified than have pedonic aspects are that (1) first priority was given to identification of soil species [4] as observed primarily at points on the landscape where individual crop plants send down roots in search of moisture, nutrients, and anchorage, and (2) configuration of land surfaces and of the soil cover on it is very complex and seems to defy analysis. Boulding has said that "the real world consists not of numbers but of shapes and sizes. It is topological rather than quantitative" (1980, p. 833). In human beings, the right cerebral hemisphere, that functions as the center for nonverbal, holistic perceptions (Nisbet, 1979) is capable of appreciating the intricate, beautiful patterns of soil bodies seen out-of-doors. On the basis of experience in the real world, the left brain develops verbal and numerical classifications, analyses and interpretations. People not only perform the extraordinary feat of learning language in earliest years, but also achieve throughout life something equally notable when they grasp and daily develop images of the native local landscape. A final achievement will be the development of pertinent and elegant numerical and graphical expressions of soil landscapes. Soil surveyors make it their business to know soil landscapes. For a century now, these people have carried in their heads fresh images of soil terrain patterns. Only a relatively small part of this lore, which was derived while the surveyors were walking and observing, is recorded specifically. Science may be said to be a "product of organized fantasy about the real world, tested constantly by an internal logic of necessity and an external public record of expectations, both realized and disappointed" (Boulding, 1980, p.832), by means of which the most valid concepts are recognized, selected out and adopted. The concept of the polypedon is one such "fantasy" that is a useful model. Testing that model against the real entities in landscape on the one hand, and against the rigors of logic (with the help of numerical analysis) on the other hand, can be expected to lead us to a better understanding of soil landscapes.

SOME FUNDAMENTAL CONCEPTS

Dokuchaev's statement (1883) that the geographic distribution of the *Chernozem* is of "prime scientific significance" is evidence that the study of soils as geographic entities dates from the earliest days of soil science. "Soil geography is mainly the study of the soil mantle, of its zonal-provincial structures, their history and causality" (Fridland, 1980, p. 642). Today, sophisticated knowledge of soilscapes resides in scientific journals, official publications, and experiences of soil surveyors. Further articulation of this knowledge is needed. To this end a framework of concepts is presented:

1. A landscape is a terrain that is distinct enough to be delineated by a knowledgeable observer.

2. Many natural landscapes have endured for thousands of years in a stable, dynamic near-equilibrium condition.
3. In a stable, dynamic landscape the biotic community or cyclic succession of biotic communities has left a characteristic impress on the soil cover of the landscape, commonly to a depth of at least one meter on uplands.
4. A terrain that is or once was occupied by a distinct biotic community or human land-management culture, the boundaries of which are approximately coincident with a pedological landscape unit, constitutes a combinational soil body.
5. Characteristics of a combinational soil body can be observed and measured. These include the number, sizes, shapes, and arrangements of component soil bodies, each of which is characterized on the basis of horizons, degree of internal homogeneity, slope, landscape position, age, and other properties and relationships.
6. Component soil bodies within a combinational soil body may have functional interrelationships with respect to drainage, movement of nutrients, and distribution of plants, animals, and human beings.
7. The present characteristics of a combinational soil body are clues to past functional relationships between its soil bodies and environmental factors, and constitute a basis for predicting its future behavior.
8. Successful programs of soil and water conservation operate within the limits of stability of a combinational soil body.
9. A hierarchy of combinational soil body units may be distinguished.

We may agree with Pitty (1979, p. 41) that "anyone with a innate geographical sense of interest and inquiry will find the soil a particularly attractive subject for study."

NOTES

[1] Terminology is from Soil Survey Staff, 1975.

[2] The authors have observed a "knife-edge" soil boundary in a road cut in Green Lake County, Wisconsin, where a Ritchey silt loam body (50 cm. deep to dolomite) met a Kidder silt loam body (200 cm. deep to dolomite) at a small cliff buried under till by a glacier about 15,000 years B.C.

[3] A general soil map of Iowa is exceptional in that it shows three kinds of soil boundaries as to distinctness (Simonson, Riecken, and Smith, 1952; Ruhe, 1969).

[4] Although trained under W.M. Davis, an observer of landscapes, C.F. Marbut focused his attention on soil profiles as he laid the foundation for soil classification and soil survey in the U.S.A.

2

Background for Soil Landscape Analysis

We now consider very briefly the contributions of several founders of soil geography, concepts of space, concepts of landscape, preliminary definitions, and ecological approach.

SEVERAL CONTRIBUTORS TO SOIL GEOGRAPHY

An achievement of modern pedology has been the documentation in the scientific literature of a "conscious recognition of the soil body, its profile and its genesis" (Jenny, 1961) as proper subjects of inquiry. "Hilgard and Dokuchaev are twin fountainheads of modern pedology" (Jenny, 1961, p. 61).*

Eugene Woldemar Hilgard (1833–1916) traveled as a youngster from his native Bavaria to a farm in Illinois, where his family settled. As a young man he returned to Europe and in 1853 he completed requirements for the Ph.D. degree at the University of Heidelberg. He held positions as State Geologist and Professor of Chemistry and Agricultural Chemistry at Oxford, Mississippi (1855–1873), Professor of Geology and Natural History at Ann Arbor, Michigan (1873–1875), and Professor of Agriculture and Botany, Director of Experiment Stations and Dean of the College of Agriculture, University of California at Berkeley (1875–1906). At a time when speculations in matters of agriculture were as varied and heated as speculations in political and religious realms, Hilgard adhered to scientific methods and displayed good judgment in interpreting soil landscapes and soil conditions. He noted that soils occur on landscapes both as distinct bodies and as transitional mixtures. He saw that native tree species assumed different forms on different soils, and from such observations he was able to predict agricultural potentials of soils. His two greatest works on soil geography are his soil map and accompanying discussion in the cotton census (Hilgard, 1880), and his book, *Soils: Their Formation, Properties, Composition, and Relation to Climate and Plant Growth* (1906).

Vasili Vasilévich Dokuchaev (1846–1903), of Smolensk, Russia, studied science at St. Petersburg University. His greatest pedological works include

*Kir'yanof (1965) recognized only Dokuchaev as founder of pedology.

Russian Chernozem (1883), and *Geological Characteristics of Soils of the Nizhni Novgorod Province* (1886). Dokuchaev developed and used in his field research concepts of the nature and genesis of soil profiles, as well as the nature and genesis of soil landscapes at both climatic (zonal) and local topographic scales. His lectures on "Our Steppes, Then and Now," included practical suggestions for combatting the effects of drought on crops. In 1890 he organized the Eighth Congress of Russian Natural Scientists and Physicians in St. Petersburg, attended by 2,000 people. His magnetic personality and enormous capacity for creative work inspired his students. His exhibits of soil collections and maps at international fairs in Paris (1889, 1890) encouraged soil research in many countries. The Dokuchaev Soil Institute became a reality after his death, nearly 40 years after he first proposed establishment of such an institute.

William Morris Davis (1850–1930), of Philadephia, Pennsylvania, published widely on the "cycles of erosion" during his long career as a professor at Harvard University (Chorley et al, 1973). He apparently decided on this research theme while reading the newly published work by T.C. Chamberlin on the geology of Wisconsin (1883). Chamberlin wrote about valleys of different ages in the Driftless Area. During his lectures Davis would cover the blackboard with elegant physiographic diagrams to illustrate concepts of landscape evolution.

One of Davis' graduate students was Curtis F. Marbut (1863–1935), who grew up in southeastern Missouri, became Professor of Geology and Physiography at the University of Missouri at Columbia, and then leader of the U.S. Soil Survey for 22 years. By translating into English a German edition of a treatise on soil classification by Glinka (1927), one of Dokuchaev's students, Marbut introduced Russian concepts into the American soil survey program. Marbut de-emphasized the deductive approach to landscape analysis used by Davis, preferring an inductive one based upon accurate description and interpretation of soil profiles. He did continue, however, to use Davis' term "mature," or "normal," to refer to well-drained soils on stable uplands. Mature soils were those zonal soils that were clearly differentiated from the original geologic materials (Krusekopf, 1943), under the influence of zonal climate and corresponding native vegetation. Among Marbut's outstanding works are *Soils of the United States* (1935), and *Soils, Their Genesis and Classification* (1951).

Roy W. Simonson (1906–) was trained at North Dakota Agricultural College and the University of Wisconsin. He was Assistant Professor of Soil Science at Iowa State College (1938–1942), and began a career with the U.S. Department of Agriculture in 1943. He became Director of Soil Classification and Correlation, U.S.S.C.S., in 1952 and continued that work until retirement in 1972, when he became Editor-in-Chief of *Geoderma*, an international scientific journal devoted to soil science and geomorphology. His duties with the U.S.D.A. took him to all 50 states, Puerto Rico, and Guam. His professional travels have also included 25 countries on all continents except Antarctica, and he has served as a consultant to government agencies in Brazil, the Netherlands, South Africa, Venezuela, and, in Rome, to the Food and Agricultural Organization of the United Nations. From his wide observations he has developed a deep understanding of the genesis of soils and landscapes, and of land use problems. Perhaps the most widely quoted of his professional papers is the one presenting a "Outline of a Theory of Soil Genesis" (1959). His paper on soil map generalization, "Soil Association Maps and Proposed Nomenclature" (1971) is of special interest for soil landscape studies.

Guy D. Smith (1907–1981), a native of Iowa, studied soil science at the Universities of Missouri and Illinois. As a soil scientist in the U.S. Department of Agriculture, he was chief architect of the new soil taxonomy (Soil Survey Staff, 1975a) and director of research programs in North Carolina, Iowa, and New Mexico (Gile and Grossman, 1979), in which three very different landscapes were carefully investigated. He taught at the University of Ghent in Belgium, and conducted field studies of soils in Trinidad, Venezuela, and New Zealand. His "Conversations in Taxonomy" with Michael Leamy have been published in *Soil Survey Horizons* (1979–1982).

Hans Jenny (1899–), from Switzerland, served as Professor of Soil Science at the University of California at Berkeley (1936–1967) after several years on the faculty at the University of Missouri. He was trained at the Swiss Federal Institute of Technology in Zurich. His *Factors of Soil Formation* (1941) and *The Soil Resource* (1980) display the originality of his work in the field and laboratory, with an ecological insight and an emphasis on processes. He has "an elegant way of connecting so many fragments of knowledge that otherwise might get lost in their various pigeonholes" (Olson, 1980, p. viii). Jenny (1980) distinguished between *vert space* (green space, above ground) and soil space.

V.M. Fridland (1919–1983) completed an academic program in geology and pedology at the University of Moscow in 1941. His graduate studies at the Dokuchaev Soil Institute, from which he received the degree of Candidate of Science in 1949, focused on an analysis of vertical zonality of soils in the Caucasus. He made subsequent pedologic investigations in the Caspian lowland, the Carpathians, forest regions of the Russian plain and other areas, both in the U.S.S.R. and abroad. For the degree of Doctor of Science in 1964 Fridland defended a thesis that dealt with crusts of weathering and with soils of the humid tropical zone. He has made many other contributions to the advancement of soil genesis and soil geography, to the understanding of the structure of the soil mantle, and to the development of soil maps, including world soil maps. His monograph, *Pattern of the Soil Cover,* (1976a) is an essential reference for soil geographers.

CONCEPTS OF SPACE

The soil is a four-dimensional space-time system (Hasse, 1968). The traditional soil genesis equation many be recast as follows:

$$SM = \begin{cases} D = f(P, R, Cl, O, AT) \\ SB = f(Pd, Rd, Cld, Od, AT) \end{cases}$$

SM signifies the soil mass, D stands for pedon, SB is a soil body, either elementary or combinational. These are functions (f) of factors: parent material (P) and parent material domain (Pd) or body of initial material; relief (R) and relief domain (Rd) or terrain configuration; climate (Cl) or climatic domain (Cld); organisms (O) or domain of organisms (Od), and space-time (AT).

Modern scientific cosmology, which takes account of relativity, assumes that Euclidean geometric statements are of only local and temporary validity (Sklar, 1974; Davies, 1977). Corrin (1953) discussed space as having three dimensions: mass, length, and time.

A consideration of space involves human beings and their images (Boulding, 1956). Fisher (1971) presented a map of "inner space," indicating degrees of awareness of the external environment. Delong (1981) reported evidence that the smaller the scale of setting in which people are active, the shorter the time intervals seem to them. The development of spatial skills in the child have been discussed by Piaget and Inhelder (1967), and by Robinson and Petchenik (1976) as proceeding by exploration of perceptual space, conceptual space, and finally, representational space. The last two authors included under the heading of cognitive elements (1) erroneous concepts, and (2) two kinds of perceptions: (a) those about milieu (environment) and (b) those about maps. The relation of space to spaciousness has been explored by Tuan (1974, 1977). Moholy-Nagy (1966, p. 3) reported that "when pious Indian Jains were asked by scandalized British missionaries why they went naked, they replied: 'But man is clothed in space . . .' "

CONCEPTS OF LANDSCAPES

The German word *die Landschaft* ("landscape") has a qualitative flavor reminiscent of *die Freundschaft* ("friendship"). Because of its many connotations, the term *landscape* is, according to Hart (1962) of doubtful usefulness in scientific work. The terms *terrain* (lay of the land) and *terrane* (an area of unique geology) are more objective, as are *soil cover* (Fridland, 1976a) and *soil body.* We may conclude, for the present, that *soil landscape,* and the abbreviation *soilscape* (Hole, 1978; Buol, Hole, and McCracken, 1980) are of value as general introductory terms, but perhaps not as specific ones. We will now consider briefly some of the connotations of the word "landscape."

Landscape is a "stretch of country as seen from a particular vantage point" (Harris, 1968, p. 627). Hartshorne (1939) states that *landscape* has two basic meanings. One is of a view or appearance of the land as people perceive it. The other is of a kind of region or restricted piece of land. In discussing the first meaning, he characterizes landscapes as the total surface, including movable and changing entities, as viewed from above. Because the soil is largely covered with biomass, in this definition it is a minor part of the landscape; this book might then be retitled: "Geographic Analysis of the Soil Resource In and Underlying the Landscape". Hartshorne was of the opinion that human activity has affected all landscapes of this planet (if we consider radioactive fall-out, this opinion is certainly true). Like Hartshorne,

Hole (1978) has emphasized the modern view of landscape as seen from an aircraft, in which the sky is eliminated from view. Tuan has been quoted by Vance (1980) as arguing for the "side view" of landscape, showing sky and land, because that is the normal way that human beings perceive the landscape.

Esthetic appreciation of landscape is a part of the human experience. The Chinese term for landscape art genre, *shan shui* ("mountain and water"), emphasizes the duality of verticality and horizontality (Tuan, 1974).

The concept of landscape as a region rests on the assumption that boundaries may be drawn separating one parcel of landscape from another. Hartshorne (1939, p. 335) quotes Bucher (1827) as saying that only arbitrary boundaries may be drawn, because the complexity of landscape is such that spatial changes in landscape properties do not coincide. At best, a cluster of boundaries based upon a number of separate landscape properties may show a zone of transition, along the middle of which a line may be drawn arbitrarily. We may conclude with Hartshorne (1939, p. 335) that landscape is a piece of area "having certain characteristics which in our minds, if not in reality, sets it off from other pieces of area."

The *Great Soviet Encyclopedia* (Prokhorov, 1973) describes landscape science as a discipline founded in the early 1900's by L.S. Berg, G.N. Vysotskii, G.F. Morozov, and other followers of V.V. Dokuchaev in Russia, and by S. Passarge in Germany. It deals with the structure, genesis, and dynamics of landscapes. ". . . geographic landscape is the primary element in the physiogeographical differentiation of the the earth" (Isachencko, 1973, Vol. 14, p. 195).

EXPLORATION OF DEFINITIONS

Soil, like water and quartz, may be regarded as an entity that cannot be subdivided and numbered (Knox, 1965). However, observations by mineralogists of varieties of quartz and discussions by limnologists and hydrologists of water lead to differentiation of units within them. So it is with soil, even though we are able to observe and sample only a small portion of the soil cover that is on the land.

In the recognition and delineation of soil bodies, emphasis is placed on relatively permanent soil characteristics, which are those that endure for decades (Schelling, 1970), rather than on variable characteristics that endure for no more than several years, and commonly for no more than a season. The dynamics of soil bodies are usually not considered in detail in soil classification and mapping procedures, whose primary goal is to record static images of soil bodies and of patterns of their assemblages.

Knox (1965), Van Wambeke (1966), Schelling (1970), and many others have clarified some of the complexities involved in identifying soil bodies by proposing certain definitions. These, along with others we offer are presented in this section. A more complete list appears in the glossary.

1. "An individual is the smallest natural body that can be defined as a thing complete in itself" (Cline, 1949, p. 81). A natural individual is the

opposite of an artificial one (see item 8 below). Anthropic (manmade) soil bodies are real and are instances of natural soil bodies in this sense.

The following components of soil are not considered soil individuals because they are not complete in themselves: (a) primary soil particles, including grains of sand, silt, and clay, as well as pebbles, cobbles, and boulders; (b) hand specimens of soil, whether pedal or *apedal;* (c) individual soil horizons; (d) two-dimensional soil profiles that exhibit the *anisotropic* nature of a soil body. A soil body is a soil individual (see item number 5 below).

2. A class is "an abstract field created by a class concept," which includes the basis for membership in it in terms of one or more differentiating simple or complex characteristics. Classes may be empty. The soil series (Soil Survey Staff, 1975a) is a class concept which determines the classification of *tesseras* (sampled volumes in pedons) in the field.

3. A universe is a superclass containing all objects and classes under consideration. A *particulate subuniverse* consists of countable non-overlapping bodies. A continuous universe is divisible only arbitrarily and the units may overlap.

4. A member-body is a body in a physical universe that qualifies for membership in a class. Where the universe is particulate, member-body individuals are fixed, clearly bounded by relatively abrupt transition zones to adjacent bodies, and are independent of observers. Where the universe is continuous, member-bodies are arbitrary and are not necessarily mutually exclusive.

5. A natural individual soil body (see item 1, above) in the soil landscape (soil cover) is a discrete, clearly-bounded body in a particulate subuniverse and is independent of the observer. Lateral boundaries are fixed by landscape reality, even though criteria for recognizing the boundaries may be arbitrary. An elementary soil body (elementary soil areal: Fridland, 1976a) is an actual natural individual with no internal pedogeographic boundaries recognizable even in the detail of research soil mapping (at scales of 1:1,000 to 1:100), in which many polypedons (bodies of soil series in the U.S.D.A. system) are subdivided.

6. A natural pedological spectral or transition area is a natural soil landscape segment in a continuous subuniverse, a subdivision of which is arbitrary. Such a transitional area may be regarded as a wide boundary or band between natural individuals.

7. The soil landscape is typically a mixed particulate and continuous universe.

8. An artificial soil individual is a human construct in a continuous universe. For example, a pedon of one square meter is an arbitrary unit, which, when designated at random on the land, establishes an artificial individual of soil. Positions of subsequent contiguous pedons are determined by the placement of the first pedon.

9. Delineated soil bodies on a map are simplified images of more or less discrete components of real soil landscape patterns. The bodies delineated on maps correspond imperfectly to those in the actual landscape.

10. Delineated soil bodies on the ground (such as those made by stakes driven into the ground) are natural geographic entities that correspond to delineations on a related soil map, in the universe of mutually exclusive soil bodies delineated on the soil map.

11. A pedon of one square meter (see item 8 above) is an arbitrary unit applied to the soil landscape, a unit that is too small to characterize soil body slope, shape, and complete soil-root relationships. Pedon siting is arbitrary and therefore pedons are not mutually exclusive.

12. A pedon of 10 square meters is less arbitrary than one that is one square meter in surface area, because the boundary of the former is precisely related to a natural pedological transition. The pedon 10 square meters in surface area is an artificially bounded transitional landscape segment (see item 6 above).

13. A tessera is an actual operational sampling volume that is usually smaller than a pedon (see item 2 above).

14. A polypedon is an arbitrary concept of a cluster of pedons judged to be similar on the basis of soil series (U.S.D.A.) criteria, and to be separable from adjacent polypedons and bodies of not-soil by boundaries which are loci of relatively abrupt change in soil properties.

15. A polypedonic soil body is an actual cluster of pedons on the ground, about 85 per cent or more of which are similar, according to criteria of a soil series (U.S.D.A.) concept, and which is distinguishable from adjacent bodies by (a) a greater change per meter of distance across the land surface at the boundary than is the case within the soil body, or by (b) an arbitrary measurement or limit in gradual change, such as depth of leached loess in a laterally thickening blanket of it overlying a uniform substratum (Figure 3.3).

16. A combinational soil body is a body consisting of two or more associated discrete elementary soil bodies (see item 5, above) and transitional areas, in a real landscape, that may be distinguished from adjacent bodies of soil and not-soil.

The interaction between the three realms of taxonomy, cartography, and pedological ecological reality is illustrated by Figure 2.1. Boulaine (1980) has similar concepts which he represents with the terms *le taxon, le mappon,* and *le phénon* (see the glossary).

OUTLINE OF AN ECOLOGICAL APPROACH

As pedologists we consider soil bodies, both elementary and combinational ones, as entities complete in themselves, that are products of interactions between material and environmental factors. The soil bodies occupy "soil space" as distinct from "vert space" (Jenny, 1980) in which green plant growth takes place. However, our awareness of the continuity of processes and organic tissues between these two spaces makes it easy for us to consider soil bodies as incomplete entities that are parts of larger wholes, namely ecosystems. This second viewpoint will now be briefly considered.

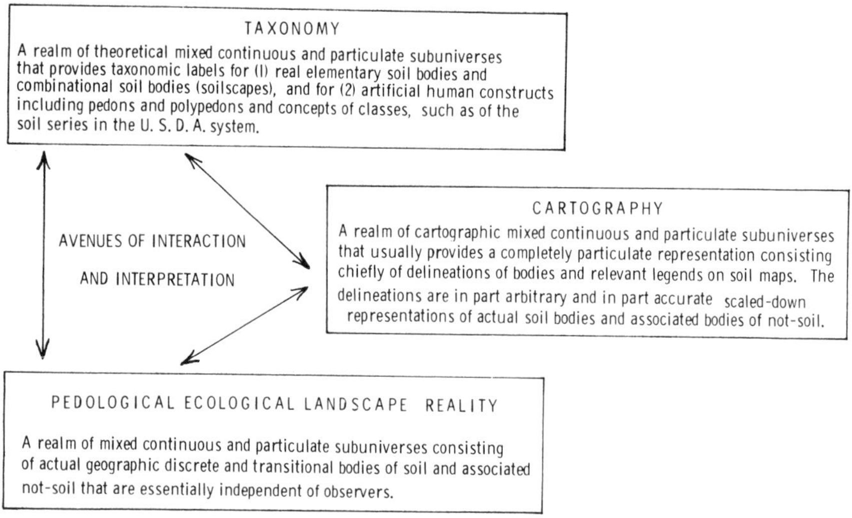

Figure 2.1. Interaction between three realms in soil landscape analysis.

Work of Forman and Godron (1981) is useful in thinking about an ecological approach to soil landscape analysis, and will form the basis of the following paragraphs. We will, however, put aside any basic limitation as to size in discussing landscapes. We prefer to start, as did Fridland (1976a), with his concept of the elementary soil body, with the assumption that ecosystems and landscapes may be of any size.

A landscape is a terrain that is characterized by repetition of a certain cluster of ecosytems. Evidence of random distribution of ecosystems is virtually absent in landscapes. Each cluster of ecosystems is also an arrangement of (1) landforms, (2) bodies of geologic materials, (3) soil bodies, with hydrologic and other ecological regimes, including (4) disturbances, both natural (such as tornadoes) and anthropic (such as harvesting of vegetation), and (5) transitions or ecotones of varying widths between ecosystems. Ecologically, landscape structure is measured by distribution of energy, nutrients and other materials, and species in relation to kinds, numbers and configurations of constituent ecosystems. It may be noted here that, although there are similarities between soil landscape and ecological landscape patterns, in a given terrain, detailed congruence of the two patterns is commonly lacking.

A landscape consists of patches that are characterized by assemblages of species (communities of organisms). The patches may be contiguous or may be set in a matrix that has different community structure and composition than the patches. The matrix of a landscape has its own characteristic degree of heterogeneity and connectivity.

Five kinds of patches may be distinguished, and these may overlap (Forman and Godron, 1981): (1) Disturbance patches are caused by localized disturbance in a matrix of relatively undisturbed ecosystems. Blow-down of trees in a small portion of a forest illustrates this kind of patch. (2) Remnant patches are relatively restricted, undisturbed ecosystems set in a matrix of disturbed ecosystems. A stand of vegetation that has escaped an extensive burn is an example. (3) An environmental resource patch is one characterized by a relatively stable local contrasting condition. Serpentine barrens, and desert oases are examples. (4) Anthropic patches are created by human activity: golf courses, cemeteries, cropped fields, tree plantations, and so on. (5) Ephemeral (quite short term) patches are produced by variations in biotic and/or abiotic conditions that come and go. Examples are intermittent seepage spots, patches on mountain slopes affected in a particular year by unusually frequent freeze-thaw cycles, and patches affected by seasonal visits of flocks of migrating birds.

Variation in the sizes of patches is associated with variation in habitat diversity, disturbance, degree of isolation, boundary discreteness, and age-related stage of vegetative succession. The edge zone of a patch is commonly a few meters to tens of meters wide. Size, shape, and spacing of patches determine the proportion of the patches that are occupied by core (if any) and by edge zone. Point-centered patches are commonly aggregated (clustered); line-centered patches (corridors) are netted, having dendritic, rectilinear, and interrupted patterns.

This outline contains many of the same elements that appear in discussions of soil landscape analysis. From an ecological standpoint a soil body functions differently when conditions change in a patch or part of a patch, such as the core or the edge zone. In this sense a soil may become a different soil within a short period of time. Baxter and Hole (1967) and Bouma and Hole (1971) found that a corridor patch of Tama silt loam (Typic Argiudoll, fine-silty mixed mesic) under prairie was quite different in many properties, including hydraulic conductivity, from an adjacent maize patch of Tama silt loam. Recently the prairie corridor in question was plowed as a part of the maize field, as a result of abandonment of a railroad line in the middle of the corridor. The dynamics of the patch of former prairie soil has been greatly altered. More attention to ecological classification of soils can be expected in the future.

3

The Nature of the Soil Cover

The *soil continuum,* called the soil cover and the soil mantle in English translations of Fridland's books (1976a, 1976b), presents a pattern that is characterized by more or less regular spatial succession of bodies of different kinds of soils. The dissimilarity of the soils creates heterogeneity locally, but the repetition of patches of local heterogeneity gives the soil a regularity of structure.

TWO APPROACHES TO THE SEARCH FOR AN ELEMENTARY UNIT OF THE SOIL COVER

It is important to define an elementary unit of the soil cover that will come as close as possible to being as definite as a "pine tree" (Knox, 1965, p. 3). One approach has been to recognize broad soil landscape units, then to subdivide them in order to define elementary units of the soil cover. In Australia, Christian and Stewart (1952) developed the "land systems" approach to soil cover description. Broad-scale landscape units are defined as composites of several smaller units more narrowly defined on the basis of soils, geology, vegetation, hydrology, and other considerations. Units are defined in a hierarchical manner, so that smaller units are nested within larger regions at several levels of detail. The original concepts were first implemented in Australia, but have since been modified for application elsewhere. Thomas (1969) suggested a six-fold hierarchical classification system for some soil regions of Africa: landform region, landform system, landform complex, unit landform, facet, and site. The U.S. Forest Service (1976) uses a six-part hierarchical land inventory system: province, section, subsection, landtype association, landtype, and landtype phase. The landtype phase refers to a parcel of terrain measuring about 0.1 km across and is the only unit of the six to which a detailed soil name is assigned (e.g., "Typic Cryochrepts, loamy skeletal mixed").

Even the evolution of a soil series in the U.S.D.A. soil survey program has been by a kind of subdivision. Figure 3.1 shows the process of refinement of the taxonomic concept of the Miami soil series in the State of Indiana from 1900 through 1977 (Sanders, Sinclair, and Galloway, 1979). The original concept of the Miami soil series was a broad one, defining a variety of soils developed on glacial drift plains in the Great Lakes Region. It was a concept of a kind of "plowland," reminiscent of the plowland types of Sibirtsev.

The modern concept of the Miami series (Figure 3.1) is based upon what is left of the original concept after the successive removal from it of portions that were used to define other soil series, including the Carrington and Tilsit series. The whole process relied upon evaluation of data from field studies of representative soil profiles and from laboratory data. By focusing attention upon soil profiles, modern workers have made enormous progress in substituting detailed morphological and taxonomic terminology (Soil Survey Staff, 1975a; 1975b) for the intuitive terminology of early naturalists (J. Bartram, 1751; W. Bartram, 1791; Carver, 1802; Hutchins, 1778; Kalm, 1770; and Strickland, 1801). The work of narrowing the definition of the Miami soil series progressively de-emphasized landscape properties and left the treatment of geographic properties in a sketchy state. The arduous process of correlation of soils, represented by this one example, has yielded more than 10,000 recognized soil series in the United States (Soil Survey Staff, 1980). Knox (1965) noted that "a polypedon is the largest possible member of a soil series," the classification of which is "independent of polypedons" (pp. 82–83). Genadiyev and Gerasimov (1980) have noted the close tie between soil series and land use considerations in the United States. The soil series concepts and the soil survey maps showing areas labeled with corresponding soil series symbols, have contributed immeasurably to the development of agriculture (Beinroth, 1978). In the U.S. results of research on soil behavior at experiment stations have been transferred to farms and ranches by noting the distributions of soils similar to those at the stations. The omission of slope and other phase considerations from the soil series category (the lowest in the taxonomy) is compensated for by the delineation on published maps of bodies of soil phases, and the tabulation of crop yields by soil slope and erosion phases in accompanying reports. Bodies of soil phases (*consociations*) are the elementary units of the soil landscape that have been discovered by this procedure. These elementary units probably lack precision insofar as (1) the major decisions in soil classification and cartography were made at the series level, and (2) two or more soil conditions were included, for convenience and practical considerations, in one soil series.

An example of the second practice is illustrated by the soil map unit, "Bearden silty clay loam, saline" (Fine-silty, frigid, Aeric Calciaquolls, somewhat poorly drained) (Doolittle et al, 1981) in a terrain in North Dakota. Bodies of this soil unit are composed of depressed (somewhat saline swale) background soil bodies occupying about 55% of the area, in which are set slight elevations (more saline swales, with a relief of about one meter) that measure 30 to 400 meters in diameter or as much as 600 meters long and a fourth as wide. From a practical point of view this combination of raised soil bodies set in a larger one is all too saline for intensive cropping to warrant formal recognition of two kinds of soil.

The first approach seems to have stopped short of defining the true elementary unit of the soil cover, assuming that there is such, probably because areal units too small to farm separately are of little interest to land managers. It is worth attempting to seek out the elementary units first and then to consider how they cluster.

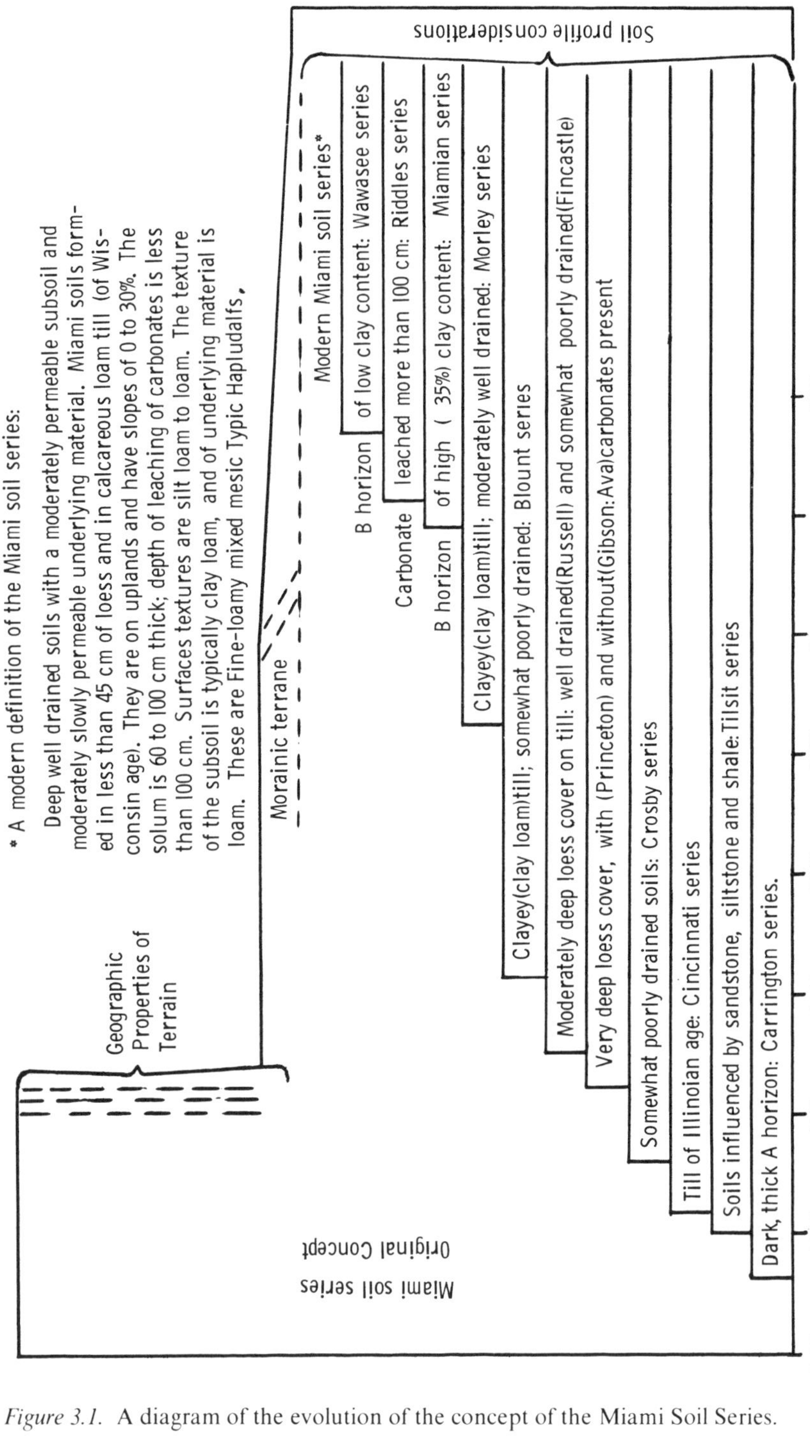

Figure 3.1. A diagram of the evolution of the concept of the Miami Soil Series.

The second approach is to start with elementary geographic soil units, the smallest individuals complete in themselves, and proceed by recognizing clusters or combinations of them in ever-broadening scope (Figure 3.2). This approach will be explored in this chapter after a consideration of some useful analogies.

SOME USEFUL CONCEPTS, DERIVED BY MEANS OF ANALOGY

Analysts of soil cover can be expected to continue to develop appropriate concepts and terms as field observations and map studies progress. However, certain analogies will probably always be of some use, including analogies from soil micromorphology, mineralogy, lithology, engineering, agronomy, plant sociology, and ecological landscape analysis.

Hole (1978) has explored some relevant concepts from soil micromorphology (Brewer, 1976) which is concerned with the plasma (mobile constituents, including clay and soluble salts), skeleton (coarse sand, gravel, stones), soil features (soil skins or argillans, concretions, etc.), and voids (spaces occupied by water and air) in bodies of soil. By analogy, the soil cover may be considered to be a mega-soil skin on the surface of the planet, coating both nonpedalogic plasma (unconsolidated geologic materials) and skeleton (consolidated bedrock) of the landscape. Bodies of water, ice, salt flats and bedrock outcrops, which from the pedonic point of view have been consistently classified as "not-soil," now become voids (with various fillings)

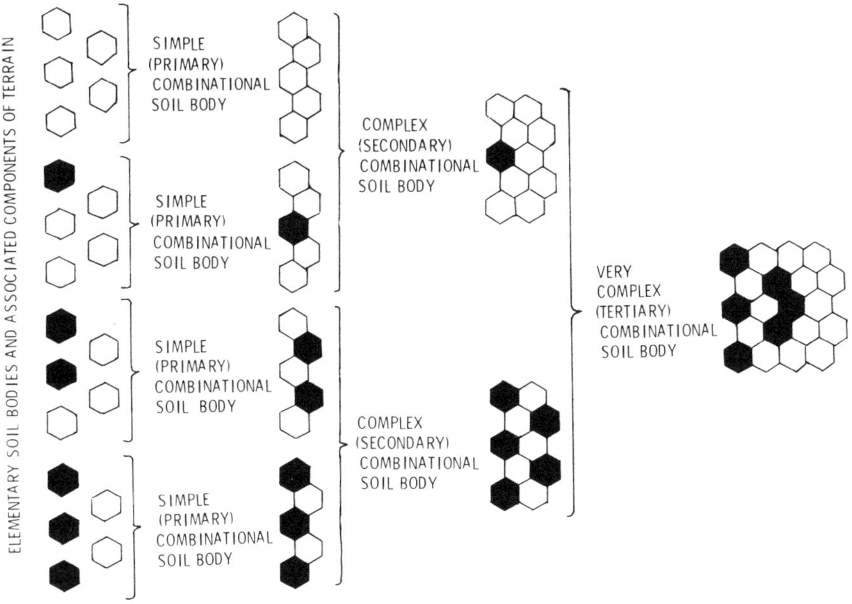

Figure 3.2. A conceptual diagram of the structure of the soil cover, conceived of as consisting of elementary soil bodies (and associated not-soil bodies) arranged in combinations of ever-increasing complexity.

in the soil cover, and these voids are integral components of the soil cover, giving it special character, as do voids in pedons.

The architects of *Soil Taxonomy* (Soil Survey Staff, 1975a, p. 3) used crystollography to form a conceptual model: "A pedon is comparable in some ways to the unit cell of a crystal." The analogy is limited insofar as unit cells of crystals are not arbitrary "artificial" individuals as are pedons (see Chapter 2, p. 15). Jenny (1965) was of the opinion that the only repetitive geographic unit in the soil landscape is the *catena*, not the pedon or tessera, since these last two do not repeat themselves as do cells in a crystal or as designs on wallpaper, but rather occur in gradient or at random.

From lithology we get the concept of bodies of rock composed of mineral grains. Rhyolite, for example, is composed of grains of feldspar, amphibole, mica, and quartz. Rhyolite may display porphyritic texture (large grains are set in a matrix of smaller ones) and may grade into or contain streaks of obsidian (natural glass). A comparison may be made to soil patterns with included water bodies. Large-scale three-dimensional aspects of lithology do not neatly apply to soil cover, which is of course very thin in relation to its areal extent.

The concept of a filter may be borrowed from the vocabulary of engineers and applied to the soil cover. We may regard the soil cover as an extensive filter through which some of the two-way and one-way flows (Figure 3.9) of aqueous solutions and suspensions occur. Portions of the soil filter may become clogged and the resulting horizons are commonly referred to as "pans."

Agronomists regard soils as media of plant root growth, and sources of plant nutrients and water. These agricultural scientists are not as concerned with theoretical genetic considerations about soil as they are with the solution of soil-related problems and with development of optimal land management practices. Zakrzewaska (1967) discussed land classifications based upon non-genetic considerations. Sanchez, Couto, and Buol (1982) have developed a technical (as opposed to natural) "fertility-capability" soil classification system that lists kinds of problems that soil landscapes may present to the land operator, problems for which solutions are available, or can be made available in a given landscape.

In phytosociology the study of plant communities includes recognition of individual plant species, then noting their tendency to cluster. Autecology deals with the study of habits of individual plant species. Synecology addresses the ensemble of plant communities, including interactions between species and the environment. By analogy, the soil geographer recognizes individual bodies of particular species of soil and proceeds to note their clusters on the landscape. Clusters or combinations of elementary soil bodies may be thought of as soil communities, particularly if interactions take place between the bodies. Plant community succession (from pioneer to climax stages) is paralleled by the much slower progress of soil community succession. *Soil autecology* and *soil synecology* might be recognized as subdisciplines of pedology.

Ecological landscape analysis provides many conceptual models (see Chapter 2, p. 12). Soil patterns are composed of disks, spots, stripes, and networks of soil entities, the dynamics of which may be viewed ecologically.

KINDS OF ELEMENTARY SOIL BODIES

There are two main kinds of elementary soil bodies: (1) discrete ones, possessing natural, definite, soil boundaries, and (2) those in transitional areas with assigned boundaries that subdivide a continuum. Figure 3.3 illustrates the difference between the two kinds of boundaries. A portion of a materials-patterned combinational soil body is shown, both in plan view and in cross-section. All bodies shown are specimens of Hapludalfs, which means that this soil landscape displays minimum taxonomic contrast. Variation in thickness of glacial till and in the kind of bedrock directly underlying the till are the major factors controlling the soil pattern. Where changes are abrupt, a sharp soil boundary occurs. Where the change is transitional, arbitrary boundaries have been drawn, based upon selected depth increments.

Fridland (1976a, 1976b) recognized three kinds of discrete elementary soil bodies (areals), which may be of any size (but rarely less than 1 sq. m in area). Figure 3.4 shows diagrams (plan view) of models of them. From left to right they are (1) homogeneous elementary soil bodies (ESBs), (2) biopatterned ESBs, and (3) regularly cyclic ESBs. In all three cases variation within the soil body is taxonomically slight. We keep in mind that biopatterns and regularly cylic patterns also exist in combinational soil bodies (CSBs). Thus, if the center body in Figure 3.4 consisted of a matrix of Rhodudults (Nitosols) with biogenic spots of Eutropepts (Cambisols), then the body would not be an ESB, but rather, a CSB in which a background elementary soil body was observed to contain contrasting smaller ESBs. Similarly if the diagram on the right (Figure 3.4) represented a body of patterned ground in which Cyorthents (simple, pale Tundra soils) occupied centers of polygons, and Cryofibrists (peats) and Cryaquents (dark, poorly drained Tundra soils) occupied borders, then the entire body would be a CSB and component bodies would be recognized as ESBs.

Homogeneous ESBs are uniform or have biopatterns with only small contrast with the background, resulting from patchiness of influence of plants and animals.

Biopatterned ESBs (see *sporadic-patchy* in glossary) have organism-induced spots. These include tree-tip mounds (bole knolls; Gaikawad and Hole, 1961) and associated depressions, small patches of mollic soils under scattered trees in a prairie matrix, and mounds made by insects and small mammals. Biopatterns are commonly considered to be ephemeral, yet some effects of plants and animals are as permanent as the effects of geologic events (Hole, 1981). For example, under long-used roosts of crows *(Corvus sp.)* a loess-derived silt loam A horizon may, in the course of centuries, be changed to a gravelly silt loam because of the habit of the birds of disgorging gravel, along with other debris, from their crops. Even in the case of

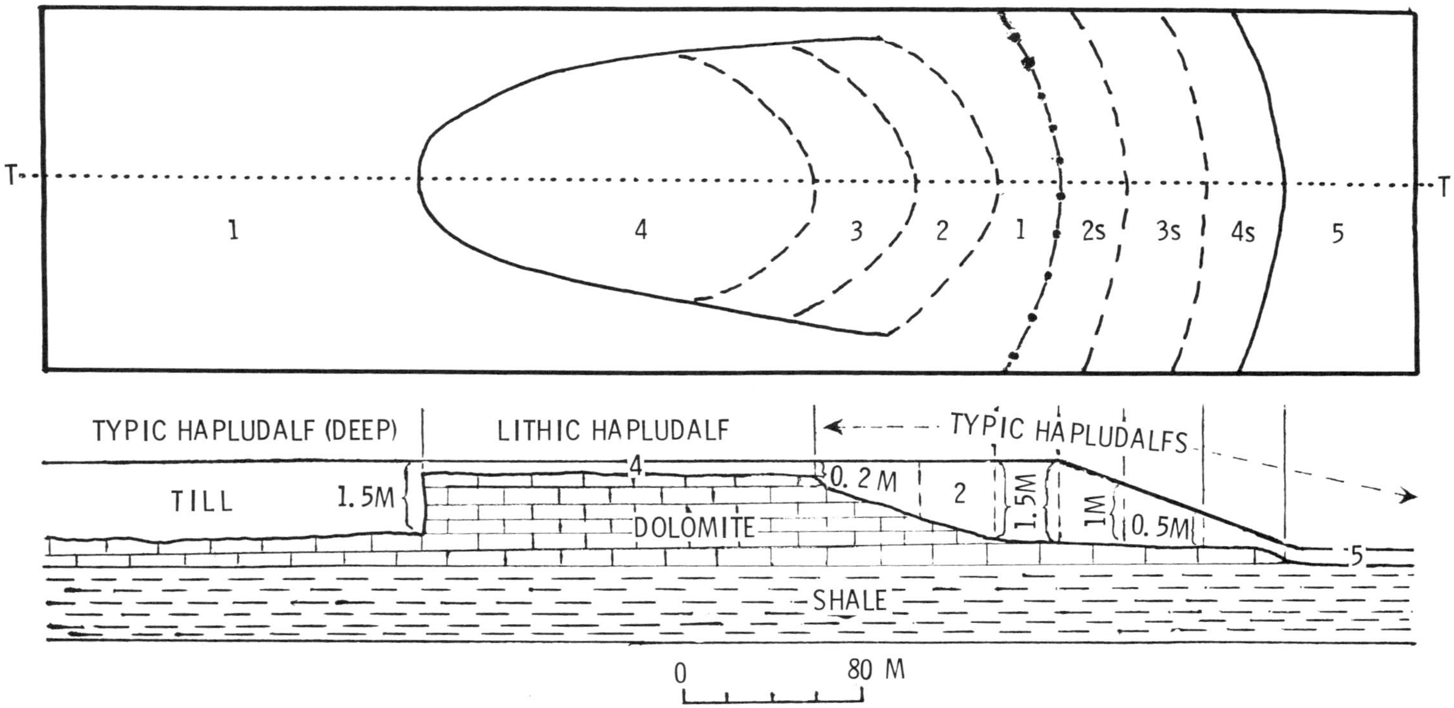

Figure 3.3. Diagram of a portion of a material-patterned combinational soil body (plan view above; cross-section below), showing natural abrupt soil boundaries (solid lines in plan view), and arbitrary soil boundaries (dashed and dashed-dotted lines in plan view) that divide wedges of loess-derived soil into segments with differing ranges of thickness. No entire discrete soil body is shown. Body number 4 comes closest to it, but is not completely enclosed by an abrupt soil boundary.

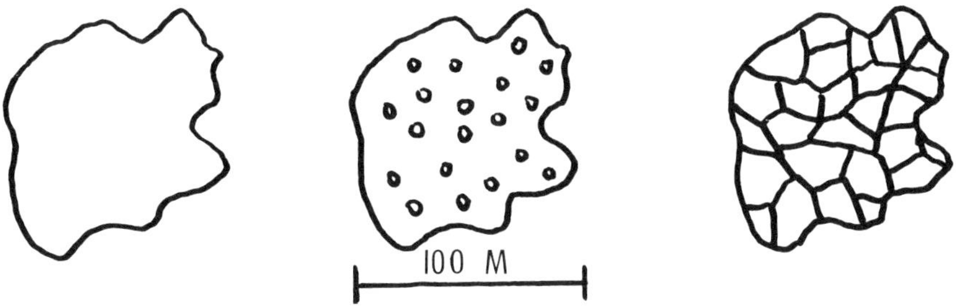

Figure 3.4. Diagrams of three kinds of elementary soil bodies, as seen in plan view. From left to right: homogenous, biopatterned, and regularly cyclic.

ephemeral bio-spots, such as the mounds of the western mound-building ant (Baxter and Hole, 1967) in undisturbed prairie, the continuous appearance of new mounds between collapsing abandoned older ones renders the biopattern itself stable and hence permanent.

Cyclically patterned elementary soil bodies are units of patterned ground, with little or no pedologic contrast, formed by frost action or shrink-swell of clay.

Elementary soil bodies are not the only components in terrains. Present also are bodies of not-soil, including bodies of water, which are usually so bonded to surrounding soil bodies by physical and biotic interchange that they are classed as active components.

KINDS OF COMBINATIONAL SOIL BODIES

Combinational soil bodies are those consisting of two or more elementary soil bodies, with or without associated bodies of not-soil. Clusters of ESBs that are free of delineated inclusions may, in a general sense (not Fridland's meaning), be called *patchwork* CSBs. CSBs that consist of relatively large background ESBs that enclose smaller ones are *punctate.* Figure 3.5 shows dozens of punctate CSBs (the included ESBs are not shown), ranging in size from about one ha to 20 sq km in Ashtabula County, Ohio. Figure 3.6 shows variations in concentric arrangement of ESBs that are arrayed beneath three soil profiles to show (left to right) even, uneven, and interrupted patterns. All of the diagrams in the figure are without designated content and therefore are pure patterns. The separate bodies of subsoil in the profile in the upper right might represent eluvial clay fillings (argillic horizons) in cavities in limestone. The pattern of interrupted elementary soil bodies in the diagram on the lower right might have resulted from shrinkage of the outermost of the three concentric soil bodies by mass-wasting processes that have moved portions of the outer soil body into the central depression over a period of thousands of years. Discontinuities in soil profiles and combinational soil bodies may be primary, that is, created by geologic agencies prior to pedologic *timezero,* or may be formed simultaneously with evolution of the soil mass.

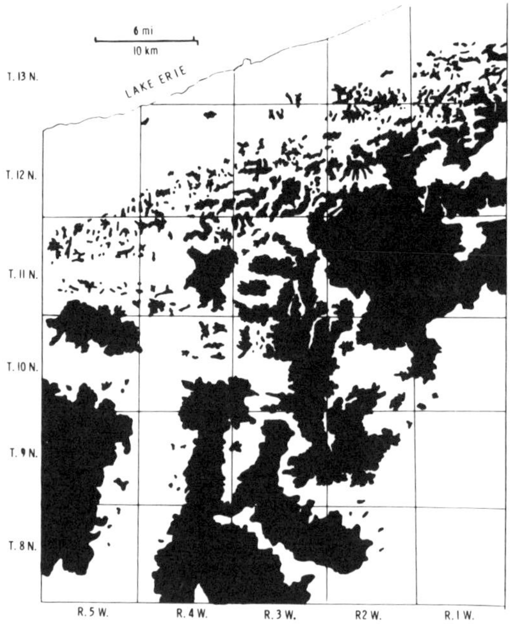

Figure 3.5. Outline map of Ashtabula County, Ohio, showing (in black) punctuate combinational soil bodies. (Rieder, Riemenschneider, and Reese, 1973.)

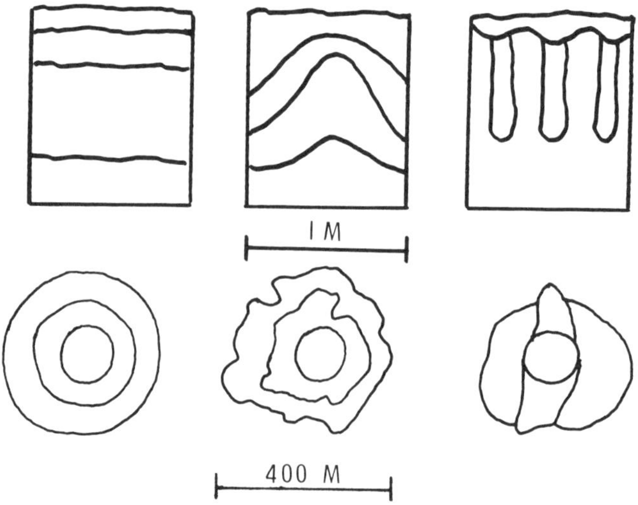

Figure 3.6. Top: Arrangements of soil horizons in soil profiles. *Left to right*: soil profiles show horizons of even thickness (within a given horizon), of variable thickness, and (in the subsoil) with discontinuity. *Bottom*: Combinational soil bodies show, left to right, elementary soil bodies of even width, of variable width, and with discontinuity.

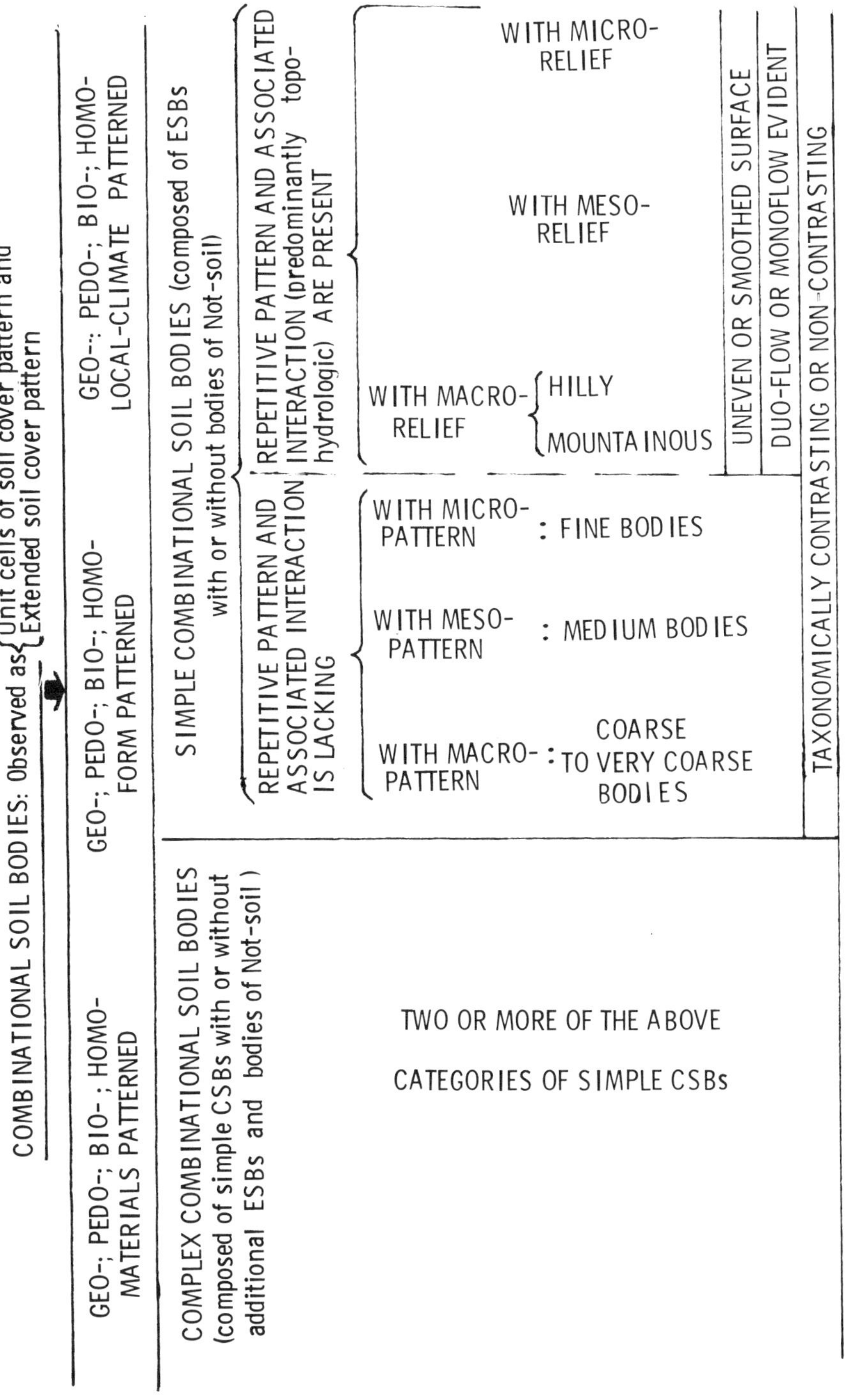

Figure 3.7 A classification of kinds of combinational soil bodies.

The great variety of kinds of CSBs is illustrated by Figure 3.7, which is considerably expanded from Fridland's classification of CSBs (see glossary: *soil areals*).

The *soil cover pattern* is defined by Fridland (1976a) as the detailed repetitive arrangement of ESBs and CSBs and associated bodies of not-soil, between which interaction (particularly the movement of moisture) is possible. Such repetition and interaction are not possible between zones and provinces, and therefore arrangements of soil zones and provinces are excluded from the concept of soil cover pattern.

Three kinds of soil cover patterns are mentioned in Figure 3.7: (a) materials patterns (which include domains of thicknesses of covers such as loess blankets), (b) form patterns, and (c) localclimate patterns.

Geomaterials patterns are the arrangements of bodies of geologic materials, such as the loess blankets just mentioned, and till inclusions in outwash. Pedomaterials patterns are arrangements of bodies of pedogenic materials and features such as clays, ironstones, and caliche. Biomaterials are arrangements of bodies of plant and animal material, peat in particular. Homomaterials patterns are arrangements of bodies of mineral and organic materials, including sythetics, resulting from human activity. Sanitary land fills, deposits of dredgings, and of fly-ash, offer possibilities for delineation of units and patterns. Geoform patterns are arrangements of landforms produced by geologic agents; volcanic craters and elevated beaches of ancient lake plains are examples. Pedoform patterns are those created by pedogenesis, such as patterned ground of tundra and gilgai. Bioform patterns are those of plant and animal origin, particularly the arrangement of mounds of various kinds. In terrains that are undergoing slow *dissection* by dendritic drainage systems, a geo-pedo-form pattern is likely to exist. The soil cover participates in the evolution of the landscape, but in a subordinate role. Even the addition of a loess blanket, which is of such consequence to soil productivity, does not much modify the onward march of the dissection. Examples of homoform patterns are arrangements of terraces, paddies, and "bogs" for production of maize, rice, and cranberries, respectively. A geo-local-climate pattern is an arrangement of soil bodies affected by local climatic domains, such as frost pockets, and relatively warm soils on xeric slopes facing the Equator. Pedo-local-climatic patterns are the result of pedogenesis. Bodies of dark soil adjacent to white salt-encrusted bodies constitute a pattern of contrasting albedo. Bio-local-climatic patterns are arrangements of soil bodies influenced by plant-animal community modification of climate: forest soils are commonly cooler than adjacent grassland soils. Thermometer birds supervise heat accumulation in mounds of litter (Hole, 1981). Homo-local-climate patterns are arrangements of soil domains whose temperature and moisture are influenced by human activity: irrigated lands, patterns of artificial drainage, the urban heat island, and barriers errected to trap pools of cool air seasonally.

It can be expected that various combinations of these categories of soil cover patterns may be recognized as research in soil landscape analysis proceeds.

A unit cell of a pattern is illustrated by Figure 1.1, which shows a CSB on a drumlin and adjacent wetland. Such units are repeated many times over in large drumlin fields (Pavlick and Hole, 1977).

Simple combinational soil bodies are composed of simple CSBs with or without additional ESBs and bodies of not-soil. There is a process-oriented or genetic bias in the classification of combinational soil bodies that is presented here. It is a bias that has been popular since the beginning of pedology, as recognized by the Soil Survey Staff (U.S.A.) in 1960: the "new concept of soil developed and introduced by the Russian school led by Dokuchaev . . . made soil science possible" (p. 1). Interaction between soil bodies in a terrain is an important part of the soil genetic process, and is given a place in analysis of soil landscapes. Although considerable attention is given to interaction, the possibility that it may be unimportant in some situations is indicated by the central part of Figure 3.7.

The three categories of local relief are defined as 0–10 m (commonly about 1 m) for microrelief (Figure 3.7), 10–100 m for mesorelief, and > 100 m for macrorelief (100–200 m for hilly terrain and > 200 m for mountainous areas) (Fridland, 1976a; Tyurin, Gerasimov, Ivanova, and Nosin, 1959).

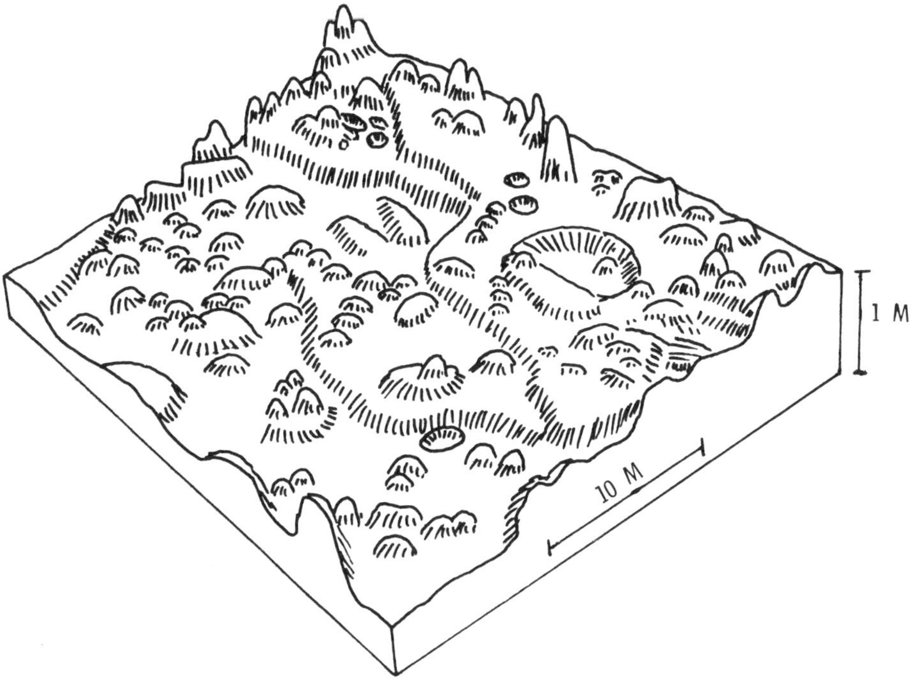

Figure 3.8. Block diagram, with vertical exaggeration, of a small portion of a soil map by Il'ina of a plot in a complex combinational soil body on the Upper Volga Lowland, U.S.S.R. (see Fridland, 1976a Figure 41). (Drawing by John Treacy.)

Although we have just defined microrelief as low rises and depressions in otherwise fairly level terrain, microrelief features may also be superimposed on moderate and steep slopes of meso- and macrorelief (Figure 3.8). Tree-tip mounds are found on a great variety of slopes, which are made uneven by the presence of mounds and associated depressions. Smoothed surfaces, on the other hand, lack such microrelief, possibly because of effects of rapid mass-wasting, or, as may be expected on nearly level to undulating terrain, because of certain kinds of agricultural operations in which rather large field equipment is used.

A consequence of bioform or geoform pattern (such as tree-tip mounded terrain, and drumlin fields) is the superimposition on the geopattern or biopattern of a hydrologically induced soil pattern resulting from runoff of water from elevations to depressions. In regions characterized by seasonal moisture deficit, "duo-flow" is important. This term refers to the movement of water down-slope, coupled with a reverse movement upward in an evapo-transpiration stream even to the crests of small knolls (Figure 3.9). In a *non-flushing regime*, precipitation is not sufficient to counteract the reverse flow of water and associated deposits of solutes and suspensoids. This phenomenon is widespread in terrains with micro-relief in semi-arid regions, with and without irrigation. The U.S. Salinity Laboratory Staff (1954) notes (p. 37) that in the western United States, where ground water is saline, a water table within 0.9 or 1.2 meters of the surface can give rise to concentrations of salts in the overlying soil horizons, including low ridges in crop rows in irrigated fields. On a salty lake plain in Utah there is an unirrigated soil complex that consists of Solonetz (Natrustalf) Leland silt loam (65% by area, all on micro-elevations), and Solonchak (Salorthid) Saltair silty clay

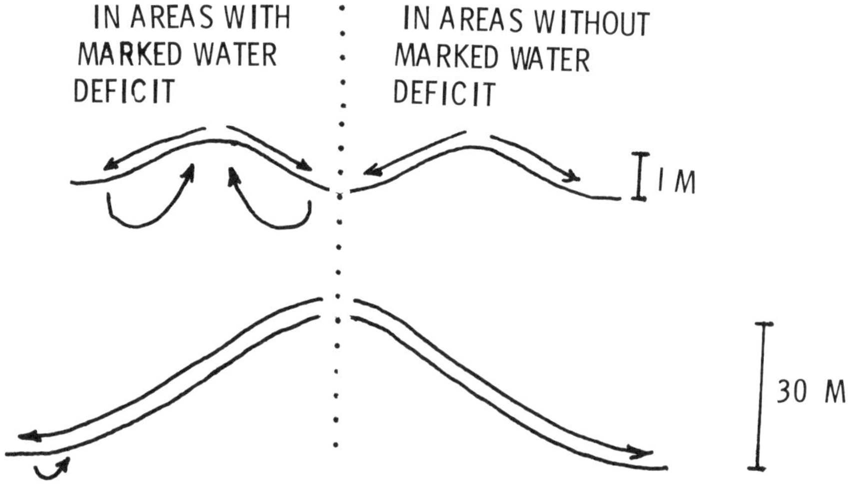

Figure 3.9. Diagram of presumed water movement above and below the soil surface, such as to leave evidence thereof, in two kinds of relief and under two kinds of regimes.

loam (30% by area in salt-encrusted depressions lying 30 cm lower) (Erikson and Wilson, 1968). It is not doubted that duo-flow occurs in such a CSB. Duo-flow is evident in the Abcal-Cache complex (Swenson et al, 1982), also in Utah, in which Cache soil, a salt-encrusted Typic Salorthid (Solonchak) is on slight elevations that occupy 30% of the area, surrounded by Abcal soil, a Typic Fluvaquent (also a Solonchak) (50% by area). Whittig and Janitzky (1963) studied a sequence of salt-affected soils with < 3m local relief in the Sacramento Valley of California, and found accumulation of Na_2CO_3 in soils immediately adjacent to a drainageway as a result of localized lateral and upward movement of bicarbonate-charged water from a low-lying anaerobic soil. In this situation only the lower part of the CSB was subject to duo-flow; the upper part was dominated by mono-flow. A reverse pattern (not accounted for in Figure 3.9) was reported by King (1982) from a salt marsh in Georgia (annual precepitation 1150 mm), in which sulphides produced by decomposition of organic materials in an anaerobic environment are diluted along tidal creeks but move up concentratedly in the interfluves that occupy 90% of the area of the marsh. In most *flushing regimes,* monoflow seems to be dominant. This is indicated by the report by Smeck and Runge (Figure 3.10) of evidence of movement of clay and phosphorous over a period of thousands of years over a distance of 43 meters to a very slight depression only 15 cm lower. Tanner and Bouma (1975) found reverse flow to be inconsequential in a flushing regime in Wisconsin. Evapotranspiration rates of 0.1 mm of water vapor per day in winter and 5 mm per day in summer from artificial septic mounds were considered ineffectual in disposing of liquid from precipitation (770 mm annually). Figure 3.9 summarizes concepts of moisture streams.

In addition to relief, Figure 3.7 mentions size of components of combinational soil bodies. Three categories are cited: CSBs composed of fine soil bodies (< 1 ha in area), those composed of medium-sized soil bodies (1 ha to 1 sq km in area), and those composed of coarse (1 km sq to 100 km sq), and very coarse (> 100 km sq in area) bodies.

The terms *contrasting* and non-contrasting refer to taxonomic diversity or its absence within a given CSB. We can say, provisionally, that if all components of a CSB belong to the same family (Soil Survey Staff, 1975a) and the categories above it, then the CSB is non-contrasting. Otherwise, the term contrastive applies. Fridland (1976a, p. 63) noted that the greater the relief of a parcel of soil cover, the greater its pedologic contrast.

LANDSCAPE POSITIONS

Elevation and slope have been shown by statistical analysis to be strongly related to many properties of soil profiles in a Tama-Muscatine soil landscape in eastern Iowa (Walker, Hall, and Protz, 1968). The late Ivar J. Nygard once said (personal communication, 1950) that the definition of classes of soil slope (A: 0–2%; B: 2–6%; C:6–12%, etc.) which are used in the legends of published county soil surveys in the United States, does not deal with the natural units of slopes in landscapes. As a regional soil correla-

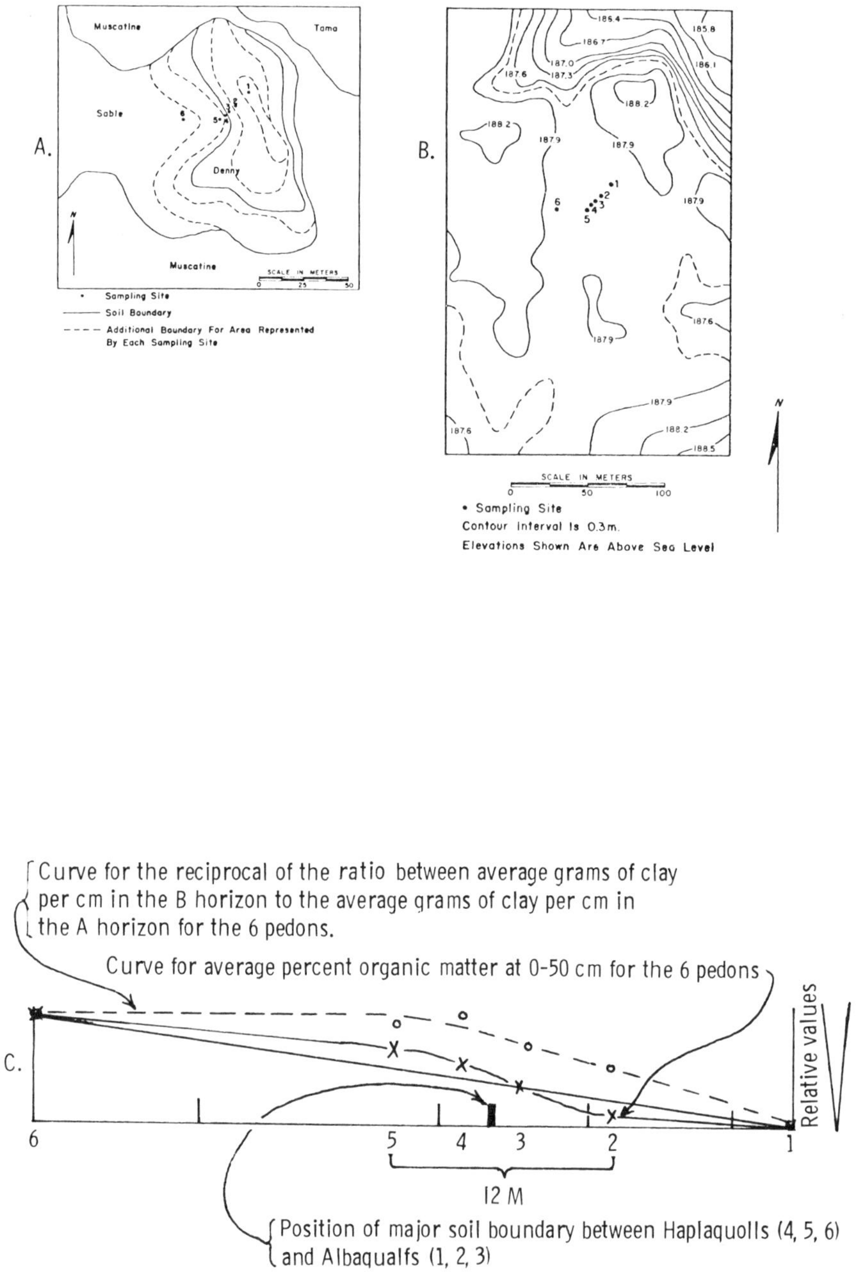

Figure 3.10. Soil map (A), Topographic map (B), and Data curves (C) indicating monoflow between two soils (an Haplaquoll and a slightly lower-lying Albaqualf) in a depression in a moraine in Illinois.(After Smeck and Runge, 1971.)

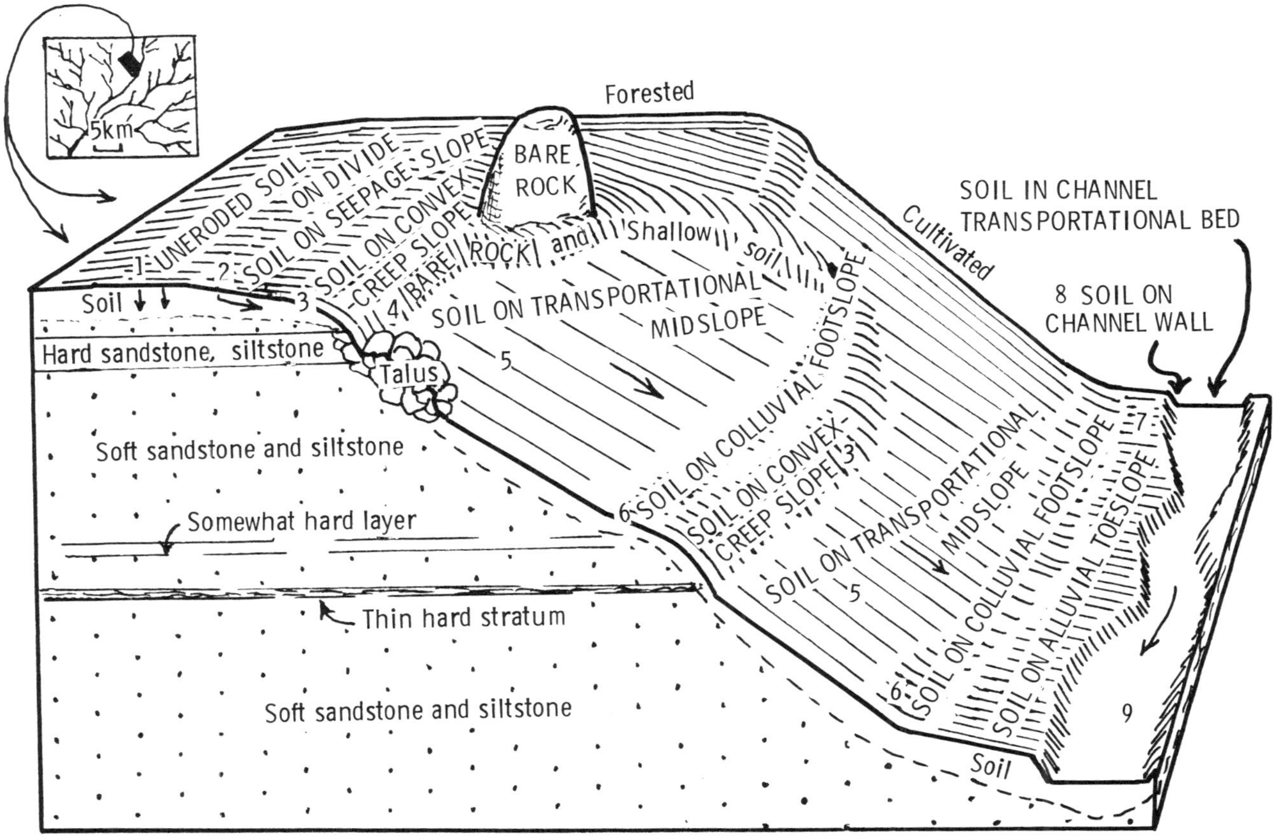

Figure 3.11. Schematic block diagram illustrating concepts of the "nine-unit landsurface model" of Conacher and Dalrymple (1977).

tor with the U.S. Soil Conservation Service, he had the opportunity to observe in Michigan, Wisconsin, Minnesota, and Alaska that slopes in many terrains consist of facets (relatively straight slopes) with breaks in slope at their junctures. The breaks between the slope classes referred to above do not coincide, in many instances, with natural slope breaks in a faceted landscape. Conacher and Dalrymple (1977) investigated this matter and proposed a generalized "nine-unit land surface model," whose units were not defined on the basis of slope gradients but rather in terms of dynamics of landscape positions. Figure 3.11 illustrates these concepts. The first position is the interfluve at which pedogenic processes operate vertically, presuming that natural drainage of the soil is good. The second unit is a gentle seepage slope that leads down from the first position, with some slow lateral eluviation of materials by water moving within the A horizon over a somewhat dense subsoil (B horizon). The third unit is called the convex *creep slope* because of the slow mass movement of soil materials across it. The fourth position is the relatively steep fall face where geologic erosion is active and weathering of fresh material is rapid. The fifth unit is the less precipitous transportational midslope, in which mass movement (creep) and surface erosion (wash) both occur. The sixth position is the colluvial footslope, where material from the upper units is deposited. Unit seven is the *alluvial toeslope* or floodplain where colluvial material is converted to alluvium by action of streams in flood stage. The stream channel wall, or slump face, is the eighth position. The ninth and last slope profile position is the stream channel bed in which material is moved both at flood and also during intrabank flows of the stream. The process of hillslope evolution is labeled DAHE in Figure 3.18.

Ruhe (1969) and Troeh (1964) have studied hillslopes in three dimensions. Figure 3.12 presents some of their concepts and terminology. Human anatomical terms ("head," "toe," etc.) as well as objective terms are used. In the profile view (Figure 3.12C), *summit, shoulder, backslope* (note that Peterson, 1981, divides this into a straight component and a concave component), *footslope,* and *toeslope* positions are designated. This is a five-unit landsurface model, in which positions are probably equivalent to Conacher and Dalrymple's first, second-third-fourth, fifth, sixth, and seventh units. In plan view (Figure 3.12A) Ruhe notes head, nose, and *side slope* positions. Troeh's concepts of water-spreading, water-gathering, convex-convex, convex-concave, concave-concave and concave-convex slopes are presented in Figure 3.12B, where the left hand portion represents a resistant layer on the interfluve as a postulated explanation for the absence of a convexity in the upper part of the profile. The right hand portion represents a situation in which rapid removal of sediment (the toeslope has engulfed and removed the footslope) has prevented the development of a concavity in the profile. In either instance, water-spreading occurs at the *nose* positions, water-gathering at *head* positions, and parallel movement of water on side slopes. One would anticipate that bodies of different kinds of soil would be found on real landscapes in each of these various landscape positions. Examination of published soil maps fails in most instances to reveal faithful adher-

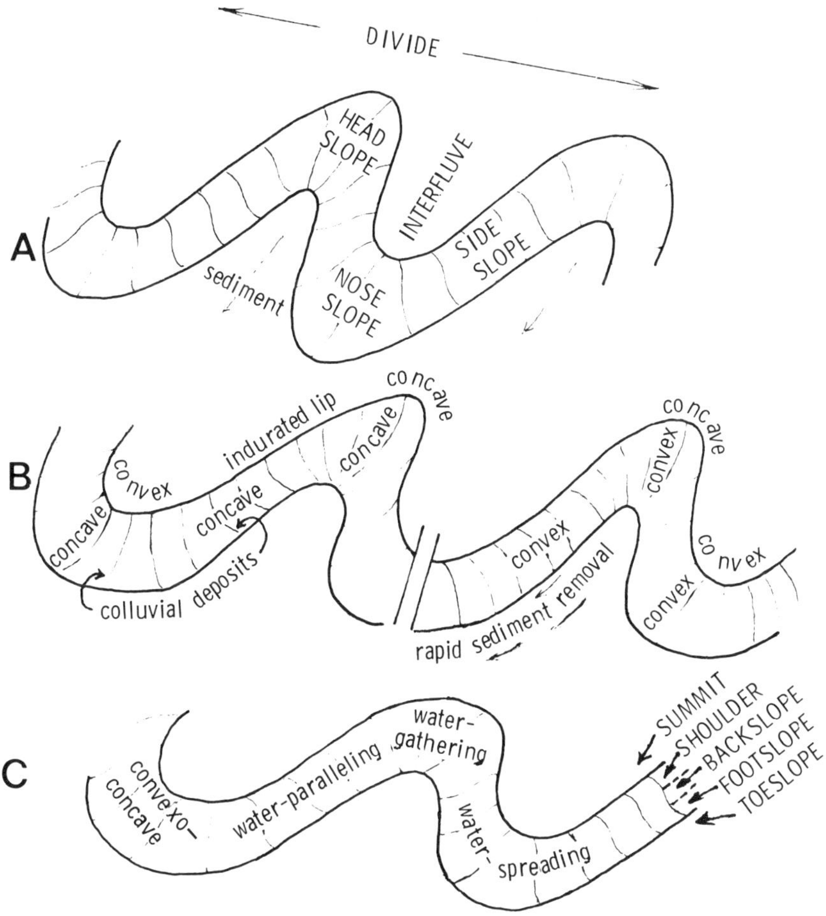

Figure 3.12. Diagrams of low escarpments, between a divide and an alluvial floodplain (above and below each, respectively) that are looped as a result of dissection by stream erosion and by mass wasting.

ence of nature to this pattern. Instead, one may find a linear soil body depicted as running along the contour (elevational line) across nose-, side-, and head-slope positions, irrespective of aspect and water-spreading or water-gathering. This suggests that materials patterns were emphasized by soil mappers and that further field observations are warranted.

Distinctions between closed and open systems are important (Ruhe, 1969). Examples of *closed combinational soil bodies* are found in inactive volcanic craters (Figure 3.17), sinks, kettle holes, and depressions between sand dunes. Soil erosion gradually leads to filling of the central depression and to reduction in internal relief. Semi-closed systems differ from closed

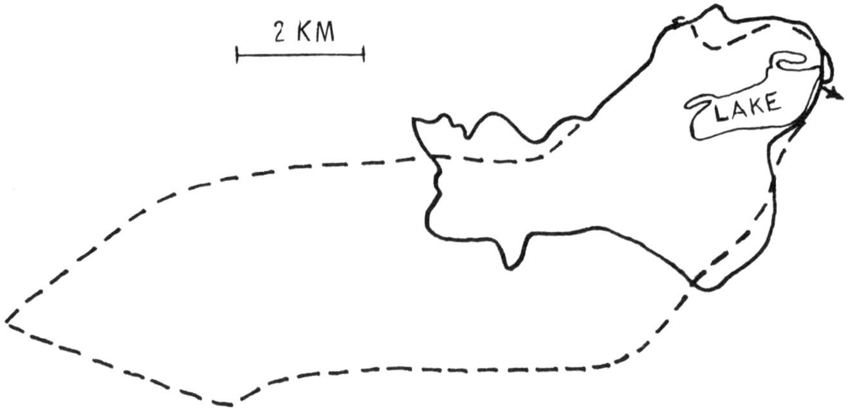

Figure 3.13. Lake Wingra drainage basin at Madison, Wisconsin. Solid line delineates the surface drainage basin; dashed line delineates the subsurface drainage basin. (Van Rooyen, 1972.)

ones by having a surface outlet. Figure 3.13 shows that one such system in Wisconsin, has a subsurface basin larger than the surface basin. *Open soil* landscapes (Figure 4.25) are those across which sediment is exported. During the process, relief may go through *Davisian* stages, increasing to a maximum, then diminishing.

Pedologists have observed that a given soil species usually has a characteristic landscape position in a particular area, but may shift to another landscape position in another geographic region (Buol, Hole, and McCracken, 1980). Characterization of landscape habits of established soil series is far from complete.

Before leaving the topic of landscape positions, we may consider for a moment the evolution of soil cover pattern. Unlike an evolving *solum* (A + B) of a soil profile, below which the initial material (C horizon) is ever present, an evolving soil body loses the initial landform. Figure 3.14 shows that the initial upland surface and well developed soil (a paleosol with O, A, and deep B horizons: stage 1) are gradually destroyed until only Entisols (OAC soils: stage 5) exist on the rapidly dissecting terrain. The mound in stage 3 is the last remnant of the paleosol of the first stage. Note that it is presumed that dissection is ecologically slow enough for the O and A horizons to be forming, even while truncation and final removal of bodies are in progress. The figure does not consider evolution of soil bodies in the adjacent lowland.

BONDING REGIMES IN COMBINATIONAL SOIL BODIES

The term *bonding* is used here to signify interaction between soil bodies. It is interaction that links many ESBs together into CSBs. Processes of interaction within CSBs are somewhat analogous to those within pedons that link horizons by eluviation, illuviation, pedoturbation (Hole, 1961) and biocy-

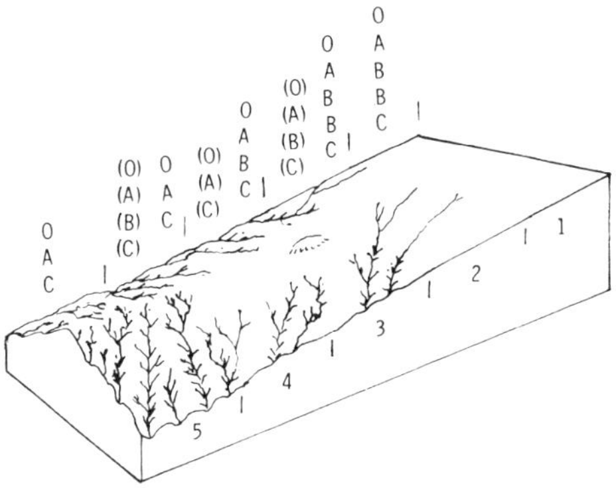

Figure 3.14 Concepts of change in soil cover showing (*left to right*) five progressive stages of dissection and evolution of an old (> 100,000 years) upland constructional surface. Letters represent soil horizon sequences of major (no parentheses) and minor (in parentheses) extent. (Adapted from Gile and Grossman, 1969, Figure 3.)

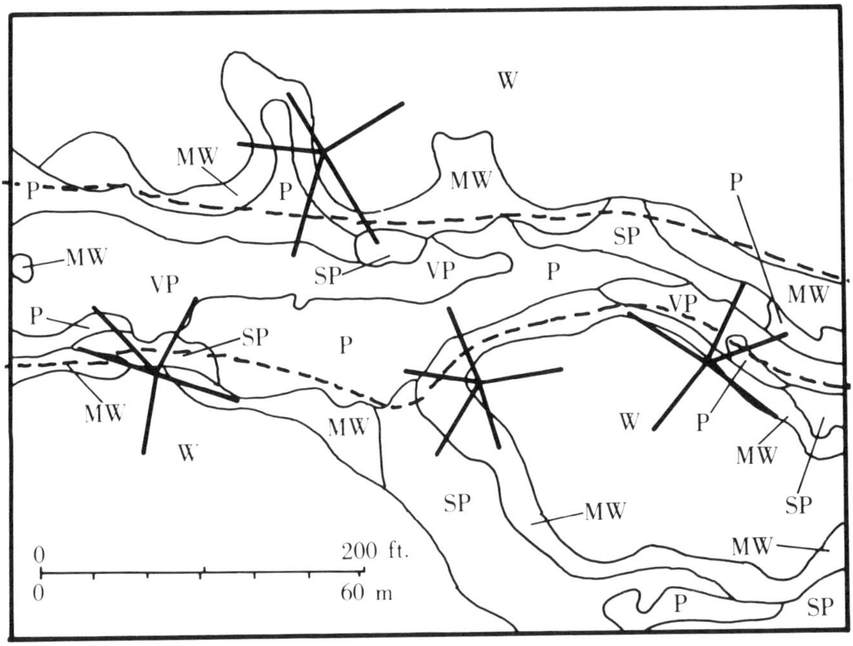

Figure 3.15 1.6 had detailed soil map of an area near Freemont, New Hampshire. Dashed lines are soil boundaries shown on published soil map (Van der Voet, 1959); solid lines, the more detailed boundaries. Clusters of radial lines (the root systems of four imaginary trees) show roots, each about 15 meters long, reaching as many as four different soils. W=well drained, MW=moderately well drained, SP=somewhat poorly drained, VP=very poorly drained soils. Scale is 1:300. (After Lyford, 1974.)

cling. Hydraulic interactions that move materials such as salts, clay, and organic matter in patterns of duo-flow and mono-flow have already been considered. Materials are also moved in and from plants and by movements of animals. Figure 3.15, adapted from Lyford (1974), shows that roots of a single tree may draw water and nutrients from four or five soil bodies and several soil moisture regimes. Ultimate decompositon of the tree will return nutrients to at least as many soil bodies. This example demonstrates one kind of biological bonding between ESBs in a CSB. Pollen and other plant particles travel many kilometers through the air and finally settle out, thus participating in bonding of widely separated soils. The phenomenon of global circulation of atmospheric dust illustrates a planetary dimension of bonding. Animal movements, both locally by worms, ants, and nesting birds, and regionally, by migration of birds, contribute to exchange between soil bodies. The extinction of the passenger pigeon (*Ectopistes migratorius*) in 1914 terminated in North America an extraordinary bond between soils of Wisconsin, and soils to the north and south. Networks of trails and drainageways are followed by many kinds of animals, whose movements contribute to the bonding of soil bodies, as do the movements of humans.

Jenny (1941) used the term *pseudoprofile* to designate a sequence of geologic material that resembles the A-B-C sequence of a soil profile. Similarly one may use the term *pseudo-bonded* CSB for a CSB with a pattern of materials arranged in a sequence that resembles an interactive CSB, yet in fact shows no evidence of true exchanges of materials.

Insofar as interaction takes place between bodies of soil and not-soil, paticularly bodies of water, the not-soil entities become bonded components of the surrounding CBS.

HETEROGENEITY OF SOIL COVER

Heterogeneity of soil cover pattern refers to numbers of soil bodies and the degree of contrast between them, on a unit area basis. Figure 3.16 shows the pattern of one section of ground moraine with drumlins in Dodge County, Wisconsin. The pattern is fairly heterogeneous, with 70 soil bodies (partial or entire) present (average area = 3.7 ha), and 11 soil series represented. Alfisols are represented in 65% of the area, Mollisols in 31%, Entisols in 2%, and Histosols in 2%. One can find on detailed published soil maps sections of land occupied by only one species of soil (see the soil survey for Ashtabula County, Ohio (Rieder, Riemenschneider, and Reese, 1973); there are also other sections displaying more heterogeneity than that of Figure 3.16.

Readers of Russian works on the *Chernozem* are impressed with the uniformity of landscapes dominated by these soils. Yet even here heterogeneity exists, as indicated by the varieties of *Chernozems* reported: unleached, poorly to well leached, meadow *Chernozems* , calcareous *Chernozems*, etc. Heterogeneity or diversity of a soil landscape may be expressed in terms of soil taxonomic complexity; mixture of entities of soil and not-soil; landscape positions, and degree and kinds, and lengths of disturbances represented (see Chapter 2, p. 15). Where superimposition of materials, bio, and hydro-

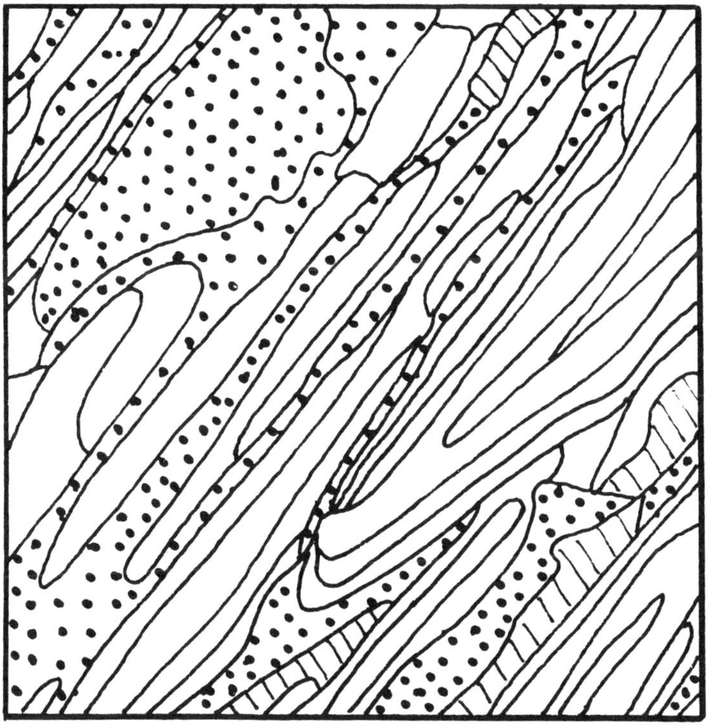

Figure 3.16 Soil cover pattern of section (one mi. sq.) 33, T. 11N, R. 13E, Dodge County, Wisconsin. Shaded elementary soil bodies have hydric moisture regime. (Fox and Lee, 1979.)

patterns occurs in intricately hilly landscapes, heterogeneity of soil cover pattern is notable indeed. This is especially the case in terrain at the juncture of several phyto-climatic zones.

FRAMEWORK OF PATTERNS OF SOIL COVER

We have progressed from consideration of the elementary soil unit of the soil cover to consideration of the framework of landforms that establish initial or parent material patterns within which soil cover patterns develop (Figure 3.18). For example, Figure 3.17 shows a generalization of the soil cover pattern in the Ngorongoro crater in East Africa (Anderson and Herlocker, 1973). The general concentric nature of the pattern was predetermined by the framework of the volcanic landform. Figure 3.16 shows a linear pattern that was established before pedogenesis by processes of erosion and deposition by an advancing continental glacier. Landforms are of differing sizes. Micro, meso, and macropatterns may be recognized (Figure 3.7). The instance of a lava flow provides a simple example of an initial landform with a definite pedologic time-zero, which is the moment at which

Figure 3.17. Generalized soil pattern in the Ngorongoro Crater. (After Anderson and Herlocker, 1973.)

the lava mass begins to cool, and environmental conditions favor development of the airborne seeds and spores that settle in crevices. Lichens and mosses commonly invade the surface. Only much later do pedologic processes play a significant role in the evolution of the landform. Pastor, Aber, and McClaugherty (1982) observed that "the distribution of (plant) communities" on forested 70 ha Blackhawk Island, Wisconsin "is a result of ecological processes which are working within the framework of a soil texture gradient determined by the geologic history of the island," involving the genesis of the landform.

Not all items in Figure 3.18 are related to major geologic frameworks and arrangements. Some ephemeral, small, biotic features are included, such as microridges made during the winter under snowdrifts by burrowing gophers (Hole, 1981). Abiotic forms (not listed) may also be ephemeral and small, as exemplified by the frost pillars, possibly no larger than a person's finger, that might appear overnight, then disappear with the morning sun.

Soil cover patterns may be characterized by a number of methods of soil landscape analysis, some of which will be considered in the next chapter.

- LANDFORMS
  - Constructional (C)
    - Abiotic (CA)
      - Deep-seated Genesis
        - Tectonic (CAT) Formed by crustal movements. Fault scarps, folds, rifts, domes. Live anticline in N. Z.
        - Volcanic (CAV)
          - Exhalative: Mud volcanoes from hot water
          - Effusive: Lava flow land forms
          - Explosive: Tephra deposits, especially ash
        - Flows (CAF) of salt, gypsum, ice water, mud forming salt glaciers, gypsum glaciers, pingoes, mud lumps. cold mud volcanoes
      - Superficial Genesis
        - Sedimentary accretion (CASA)
          - Linear: dunes, drumlins, beaches, eskers
          - Triangular: fans, deltas.
          - Extensive blankets
            - With point sources: annular, plume-shaped ash falls
            - With linear sources: Loess blankets, Glacial drift sheets, Ash falls, wash plain deposits
    - Biotic (CB)
      - (By concentration and storage of diffuse energy)
        - Mounds
          - By plants (CBP): Kauri tree "basket Podzols"
          - By animals (CBA): Mounds by thermomenter birds; by fossorial animals; of guano formed by birds, bats
        - Ridges: Built by gophers in alpine landscapes
  - Degradational (D)
    - Abiotic (DA)
      - Multibasinal: by solution or melting
        - Karst in carbonate rock (DAKC)
        - Karst in gypsum (DAKG)
        - Karst in salt (DAKS)
        - Karst in glaciers (DAKGL) and frozen ground (DAKF)
      - By mass-wasting (slump, slide, creep) (DAMW)
      - By hill-slope evolution (DAHE)
      - By fluvial erosion (DAFE): Flood plains, levees, backswamps, point bars, cut banks, dentric, trellis, annular, radial and structurally-controlled fluvial systems
      - By deep weathering processes in savanna, intertropical (DADW) (Inselbergs)
      - By periglacial processes (solifluction) (DAPP)
      - By glacial processes (DAGP) Drumlins. Roches moutonnées
      - By coastal processes (DACP) Cliffs, bars, beaches
      - By meteorite impact (DAMET)
    - Biotic (DB)
      - By plant-induced weathering, loosening, erosion (DBP)
      - By animal-induced loosening, transport (DBA)

Figure 3.18. Framework of landforms and patterns of terrain on and within which pedogenic soil cover patterns develop.

4

Methods of Soil Landscape Analysis

Methods of collecting data, both descriptive and numerical, for the characterization of soil cover patterns are applied both in the field and in the cartographic laboratory. Fridland (1976a, 1976b) quite properly commented on the need for more research and development of methods of soil landscape analysis. Each kind of landscape may be expected to dictate a special emphasis in analysis. The material presented in this chapter may be looked upon as a kind of primer on the subject and as a prelude to exhaustive studies that will be made in the future with the help of computers. Exploratory work using computers in statistical analysis were made for certain terrains in Wisconsin by Pavlick and Hole (1977), who used discriminant analysis to study one sq. km quadrats of a detailed soil map of a till plain (Fox and Lee, 1980). Nwadialo (1978) used autocorrelation in analyzing data from samples taken along catenal transects. Whatever the degree of sophistication of statistical analysis, common sense is important in soil landscape studies.

A great number of published soil maps are available to the public on library shelves, and many more are in the files available to researchers. In the United States alone this backlog of information about soil landscapes could keep many analysts busy for decades. Analysis of published maps should proceed with caution. Experienced soil surveyors are best qualified for the work, particularly those who have walked the terrains under study. Maps differ from county to county, state to state, nation to nation, and from decade to decade. The scale of a map, and the nature of the cartographic base are variables that are especially important, and meaningful comparisons between different soil cover patterns are most readily made within the confines of a single published survey. In any case, analysis of published soil maps will provide ample justification for additional field work.

FIELD METHODS

Field data are collected by (1) mapping of soils, using appropriate legends, scales, and cartographic bases (including aerial photography and similar images); (2) detailed observation and sampling of soils at points (pits, bore holes), and along lines (transects, trenches); (3) measurement of fluxes (of energy, gasses, water, and other materials); and (4) recording of phenological events of both pedological and biological significance. To this list may be added the making of a topographic map, particularly in research mapping of soils (a contour interval of 10 cm may be justified). These observations may be carried out simply by a small farmer (Collins, 1972) or by a complex procedure administered by a large agency.

Mapping Soils and Observing Soils Along Transects and in Pits. Routine soil mapping has changed in the eighty or ninety years of soil survey. For example, map scale in northern Wisconsin, has changed from 1:500,000 in the first decade of this century to 1:15,840 at present, and, in research mapping, to 1:300. Early maps were drawn on white paper and published with black lines and symbols on a color pattern. Today soil surveyors record field observations on remotely sensed images, and published maps now represent soil patterns using black lines and symbols printed on a photographic base. Current developments will probably result in further changes in the appearance and geometry of cartographic representations of the soil landscape.

Map legends have expanded as classification of soils has been refined to include a greater number of soil species recognized as part of the soil cover pattern. The trend from coarse to detailed mapping parallels the trend in classification indicated in Figure 3.1. Recent research mapping has initiated a complementary approach, that of characterizing soil cover pattern in ultra-detail in representative plots. This reveals the actual content of more generalized maps, and provides information on which to base the grouping of simple combinational soil bodies into more complex ones (Figure 3.2). Field methods of soil survey are described in the *Soil Survey Manual* (Soil Survey Staff, 1951; 1975b).

Examples of research soil maps include those by Il'ina (Figure 41, Fridland, 1976a; see Figure 3.8 of this book) of a plot in the upper Volga lowland; by Smeck and Runge of a till plain in Illinois (Figure 3.10); by Lyford of a plot in New England (Figure 3.15); by Mace of a plot in Wisconsin (Figure 4.1), and by Campbell of a plot in Kansas (Figure 4.2).

Mace (1980) surveyed approximately 35 ha of soil cover developed in glacial drift terrain in northeastern Wisconsin (Figure 4.1). He enhanced remotely sensed images by spectral classification and digital reflectance smoothing to produce a color image of the terrain that showed areas of well drained soils in bright colors (yellows and reds) and of poorly drained soils in dark colors (greens, blues, purples). With the help of an experienced soil scientist he classified soils as observed in numerous pits and bore holes (as closely spaced as 30 m), and drew many more boundaries (Figure 4.1B) than are shown on the previously published soil map (Figure 4.1A; Mitchell, 1980). The increase in the number of mapped soil bodies from 11 to 101 was

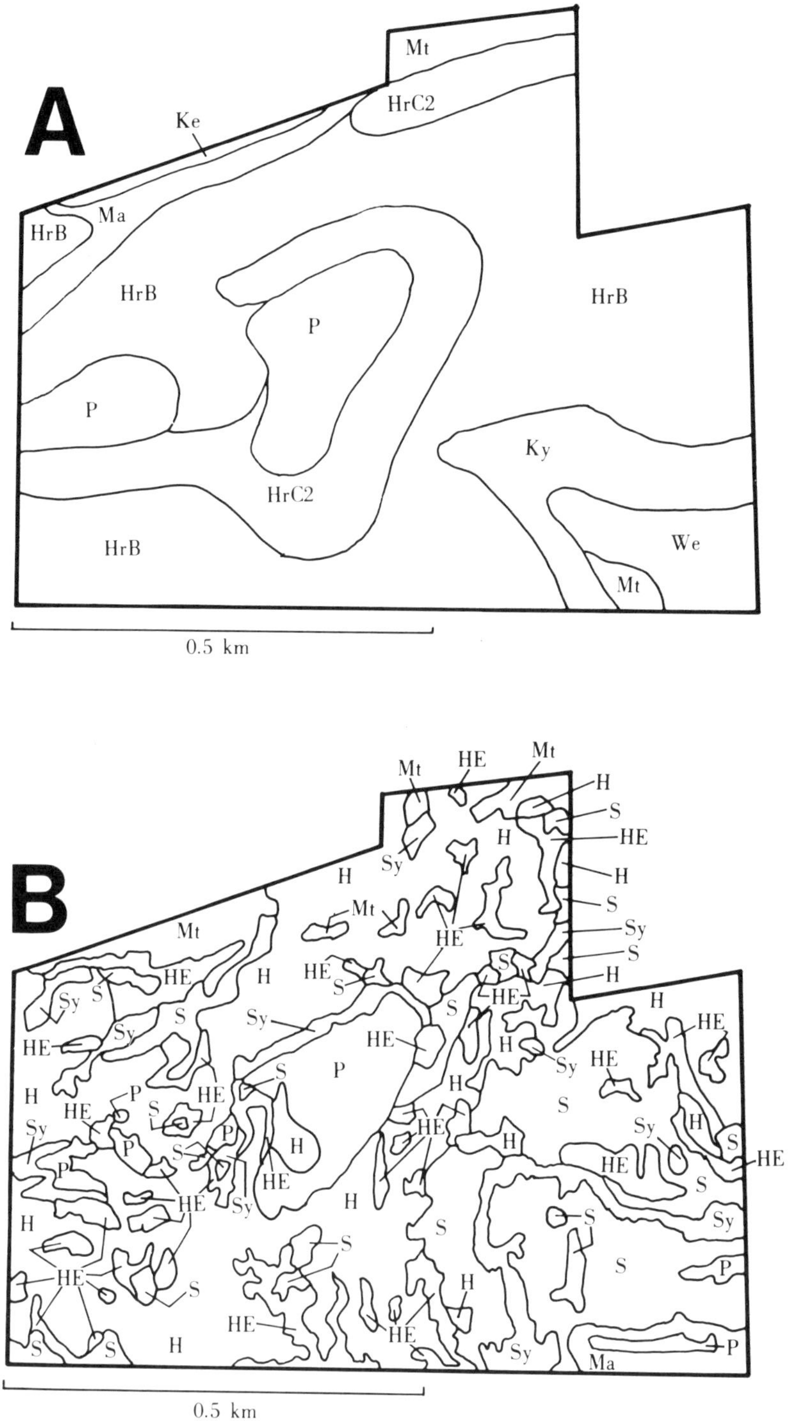
A
Mt
HrC2
Ke
Ma
HrB
P
Ky
We
0.5 km
B
Mt
HE
H
S
Sy
P
Ma
0.5 km

Figure 4.1 Two soil maps of the sames 50 ha area in Winnebago County, Wisconsin (Sec. 33, T 20 N, R 16 E). A. Published soil map. B. Research soil map, based upon interpretation of color infrared aerial photography and field investigation (After Mace, 1980 and Mitchell, 1980.)

H	Hortonville silt loam, eroded
HrB	Hortonville silt loam, 2 to 6% slopes
HrC2	Hortonville silt loam, 6 to 12% slopes
HRE	Hortonville silt loam, severely eroded
Ke	Keowns silt loam
Ky	Korobago silt loam
Ma	Manawa silty clay loam, 0 to 3% slopes
Mt	Mosel silt loam, 0 to 3% slopes
P	Poygan silty clay loam
Sy	Symco silt loam
S	Symco silt loam, eroded
S+	Symco silt loam with overwash deposit
We	Wauseon silt loam

Hortonville is a well-drained soil, fine-loamy, mixed Glossoboric Hapludalf.
Keowns is somewhat poorly drained coarse-loamy, mixed, mesic, nonacid Aquollic Hapludalf.
Korobago is somewhat poorly drained coarse-loamy over clayey, mixed, mesic, Aquic Eutrochrept.
Manawa is somewhat poorly drained fine, mixed, mesic Aquollic Hapludalf.
Mosel is somewhat poorly drained fine-loamy, mixed, Aquollic Hapludalf.
Poygan is poorly drained fine, mixed, mesic, Typic Haplaquoll.
Symco is somewhat poorly drained fine-loamy, mixed, mesic, Aquollic Hapludalf.
Wauseon is poorly drained coarse-loamy over clayey, mixed, mesic, Typic Haplaquoll.

made possible not only by the greater number of field observations, but especially by the highly detailed base map, in color.

By sampling soil pedons at 10 m intervals on a grid in a landscape in Kansas, and analyzing samples for pH and sand content, Campbell produced a detailed delineation of a soil boundary (Figure 4.2).

Nwadialo (1978) took soil samples at a depth of 30 cm at 50 cm intervals along a 75 m catenal transect in each of two glaciated terrains in southern (Dane County) and northern (Florence County) Wisconsin, and completed particle size analysis (Figure 4.3: domains 1 through 9 for Dane County; domains I through V for Florence County). The plotting of data points on the textural triangle of that figure show that conditions were more uniform in the northern soilscape (lower right in the diagram) than in the southern one. The transect at the Dane County site began in a pedon formed in deep loess on a footslope (domain number 1, right), proceeded to a shallow *solum* in glacial till at midslope (domain number 6), and ended in a pedon at the summit position (domain number 9), where the leached loess is only moderately deep. At these three sites, the samples (at 30 cm) came from the A horizon (in leached loess), from a weak B horizon (in sandy loam till), and in the upper B horizon (well developed in loess), respectively. The

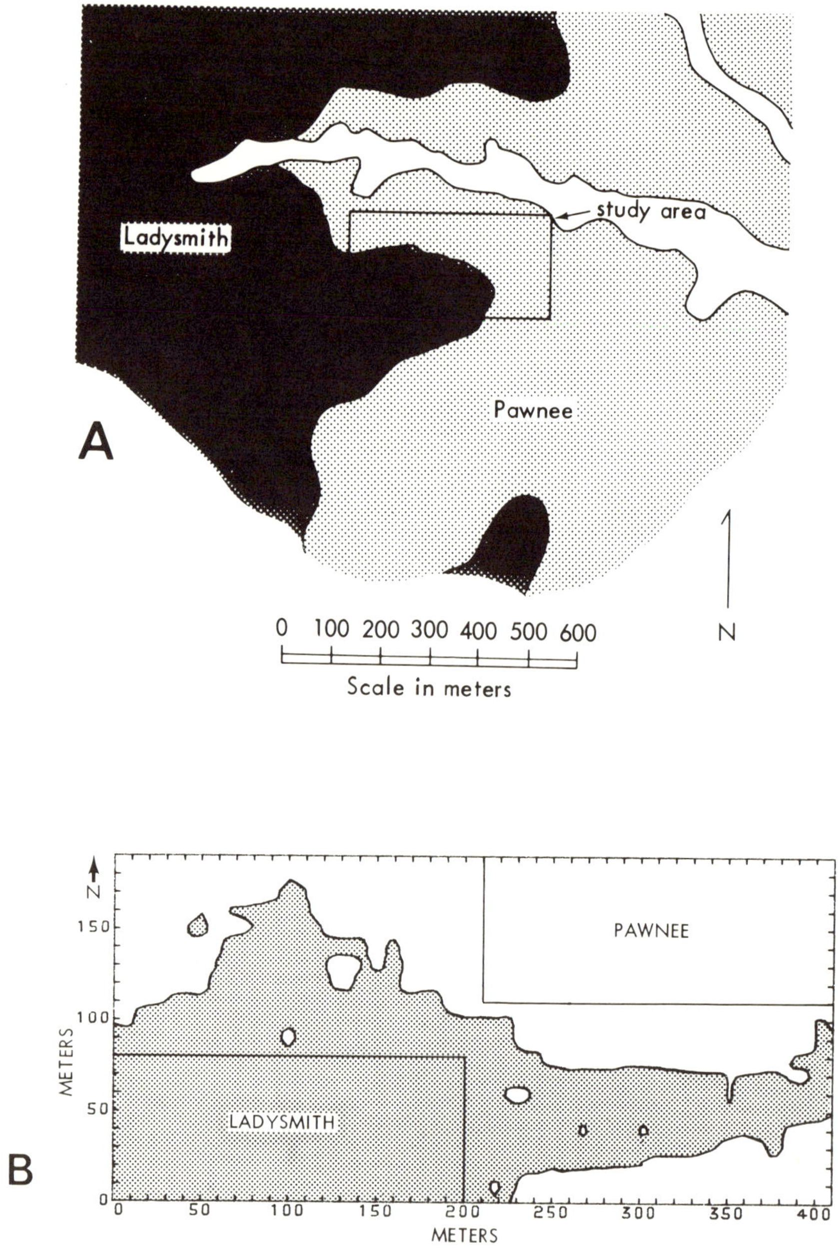

Figure 4.2. Two soil maps of the same portion of Shawnee County, Kansas (Campbell, 1978). A: Map from sheet published at 1:20,000 (W. Abmeyer and H.V. Campbell, 1970). B: Detailed soil map by J.B. Campbell (1978) made at a scale of 1:500, based upon auger borings spaced at 10 m intervals on a grid.

sequential plotting of data points, however, revealed a cluster pattern. Boundaries were drawn around the clusters (called "domains" in the figure). The number of "back-trackings" of the data sequence from one domain to another was recorded in light numerals between domains. Where back-tracking did not occur, the numeral "0" was used, and a relatively sharp boundary was indicated by dashed lines. Three of these appear in the figure, separating four textural regions, which are thought to correspond to soil bodies on the hillslope. The published soil map (Glocker and Patzer, 1978) shows a single boundary, separating only two soil bodies. Permanent soil study pits are located in the three soil bodies represented by domains 1, 6, and 9. This exploratory study illustrates the use of a limited number of kinds of data when a great number of samples must be processed.

Mockma, Whiteside, and Schneider (1972) ingeniously transferred boundaries of county soil maps (at scales of 1:63,360 to 1:48,000) published in the 1930's to aerial photographs (1:15,840) and updated the legends on the basis of 30 to 60 field observations in bodies of each mapping unit. The observations were made at 330-ft (100 m) intervals along transects positioned at least 660-ft (200 m) apart.

"Though aided in many areas by mechanical probes, transect work remains relatively slow and tedious," wrote Doolittle (1982) in a paper describing the application of ground-penetrating radar, which made unnecessary much probing and digging. A radar antenna towed across the soil surface recorded an almost continuous record of variations in thickness and character of surface and even subsurface soil horizons, along lines between auger holes and pits in the centers of elementary soil bodies.

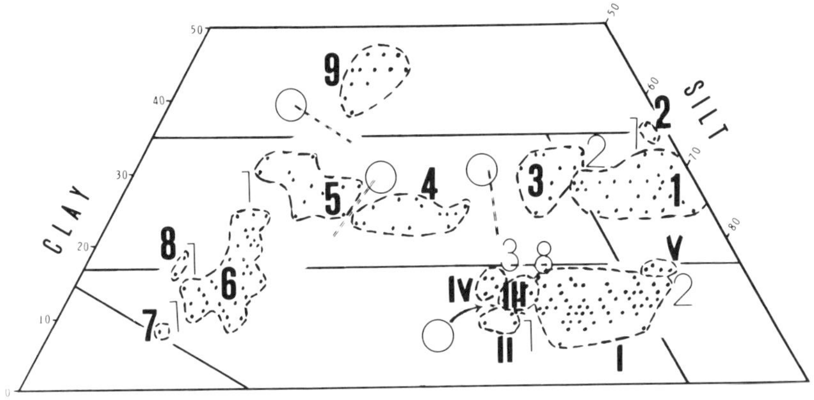

Figure 4.3. Plot of particle size distribution analyses of soil samples taken at a depth of 30 cm and at 50 cm intervals along two transects 75 meters in length in Dane and Florence Counties, Wisconsin, respectively. Dots represent data points. Dashed lines are estimated boundaries of domains or clusters of points with bold arabic numerals for Dane County data and Roman numerals for Florence County data. Thin arabic numerals from 0 through 4 represent numbers of recrossings (see text) between domains, or definiteness of soil boundaries, which are assigned lower numbers as boundaries become more abrupt. (Nwadialo, 1978.)

Some workers have noted that aerial photographs taken from low-flying aircraft at appropriate time intervals after a rain, show detailed soil patterns in cultivated fields. Milfred and Kiefer (1976) found that one day after a 2.5 cm rain on a 30 ha field of very young maize, the field displayed an almost uniformly dark tone; but two days later an intricate pattern of light and dark colors was created by differential drying of crests of micro-knolls, their slopes, and intervening depressions. An almost identical pattern was later displayed as variations in colors of the maturing maize plants themselves, in response to geographic gradients in moisture stress in late summer. The maize crop was being used as a "ground-penetrating" recording system for soil pattern delineation. The published soil map of the area (Glocker and Patzer, 1980) shows four ESBs (whole or partial). The research map by Milfred and Kiefer, based upon their photographs and upon supporting data obtained by probings of soil bodies in the field, revealed 27 soil bodies. Cooper (1982), analyzing a less adequate array of imagery (Landsat) for the Van Buren Ranger District in Missouri (43,115 ha), was unable to observe soil boundary traces in images of moisture stress in forest canopy. Olson (1981) described sequential testing of agricultural crops as a means of discovering variations in the suitability of soils for certain crops. The method also reveals patterns of soil cover. Distribution of yields of alfalfa that was planted across a contrasting soil pattern corresponded to the soil pattern.

Some Other Kinds of Field Measurements. For each soil map unit, soil profile descriptions from at least three representative sites, and representative laboratory data from carefully collected samples are essential (Soil Survey Staff, 1975a; Ciolkosz et al, 1982).

Phenological information is quite useful. This may include: (1) quantitative analysis of gasses emitted from soils of different ecosystems during the march of the seasons; (2) temperature changes through space and time; (3) changes in the content and tension of moisture through space and time, from which directions and rates of movement of soil water may be determined; (4) measurement of rate of vertical and horizontal movement of coarse fragments–and of entire soil horizons and sola of soil–and of rate of erosion of soil domains; (5) seasonal changes in character and thickness of the surface litter layer (O horizons), populations of micro-fauna (including nematodes, acarina, and collembola), cracking patterns of surface soil, appearance and disappearance of frost pillars, and patterns of microrelief.

Integrated Surveys. An integrated survey of a terrain is one conducted by an interdisciplinary team of environmental scientists. Such surveys are rare. An effort in this direction was made by Milfred et al (1967), in Menominee Indian Tribal Lands in Wisconsin (Hole, 1975).

LABORATORY METHODS

Let us assume that we have one or more soil maps before us and that we are ready to analyze them. We are most fortunate if we are working with large

Table 4.1 Outline for the Analysis of the Pattern of an Area of Soil Cover

I. Setting.
- A. Materials pattern (geo-, pedo-, bio-, homo-)
- B. Form pattern (geo-, pedo-, bio-, homo-)
- C. Micro-climate pattern (geo-, pedo-, bio-, homo-)

II. Scale.
- A. Local relief classes
- B. Degree of smoothing
- C. Sizes (areas) of elementary soil bodies

III. Principal kinds of patterns (plan view)
- A. Undifferentiated (solid) in a given quadrat.
- B. Differentiated
  - Without alignment
    - Patchwork
    - Punctuate
      - With solid background
      - With patchwork background
  - With alignment
    - Point-centered (concentric)
    - Line-centered
      - Unidirectional (striped)
      - Multidirectional (looped, etc.)

IV. Origin of soil cover pattern
- A. Geologic at "time-zero" of the pattern
- B. Geo-pedo-local-climo-biotic, during evolution of the pattern after "time zero" of the pattern
  1. Constructional: accumulation of materials in progress
  2. Degradational: removal of materials in progress: Multibasinal, Stream dissectional, By mass-wasting, Aeolian transportational, Hillslope formational (horizonational by eluviation – illuviation; catenal by geographic translocation associated with mono- and duo-flows)
  3. Over-printing by biotic communities

V. Population of soil bodies per unit area: measured by
- A. Count of bodies (entire and partial)
- B. Count of mean number of soil boundaries intersected by a transect of unit length
- C. Ratio between the count of soil boundaries to a comparable count of number of contour lines (at 3 m or 10 ft interval)
- D. Count of nodes (junctions of soil boundaries) per unit area

VI. Composition
- A. List of taxonomic units (pedota) and not-soil units
- B. Proportionate extents of components

VII. Soil cover diversity, expressed in terms of: Taxonomic or soil characteristic contrast, soil body size distribution, number of soil map legend units per unit area, number of soil landscape positions, soil moisture regime diversity (range, number of regimes, index of mean condition, index of mean condition times a climatic effective moisture index), soil body shape index, degree of distinctness of soil boundaries, taxonomic rank of soil boundaries, degree of chronological uniformity, degree of openness, length of drainage ways per unit area, pedologic sequences along drainage ways and normal to them, degrees of complication and simplification by human activity, degree of bonding of the soil cover.

scale maps made consistently over an area of at least 2,000 sq km within a four year period by a single team of experienced soil scientists. Table 4.1 presents a list of items that may be treated in the course of analysis (see also Figure 3.7).

I. *Setting*: A summary statement is useful in presenting information obtained from environmental surveys of a given area. This may include the items discussed in Chapter 3: materials patterns, form patterns, and local climatic patterns.

II. *Scale*: The local relief classes are defined in the preceding chapter. Published topographic maps and aerial photographs are referred to for areas not visited by the soil landscape analyst. No horizontal distance nor standard area is stipulated here as defining the meaning of "local." In order to judge the degree of smoothing of a parcel of land, the analyst needs to have access to information about remnants of terrain that have not been smoothed (Gaikawad and Hole, 1961; Bouma and Hole, 1971).

Size (area) of elementary soil bodies (ESBs) may be reported to indicate horizontal scale of entities. The soil body size distribution triangle (Figure 4.4) presents a useful classification and shows that data from a given terrain may be reported on the basis of total number of soil bodies (disk-shaped symbols in the figure) or of total area of soil bodies (rectangular symbols). The first basis commonly yields the smaller values. Solid squares and disks represent data points derived from published soil map sheets with a scale of 1:15,840 (Habermann and Hole, 1980). Hollow squares and disks are derived from map sheets with scales of 1:20,000 and 1:24,000. The asterisk at the top of the triangle represents some 140,000 elementary soil bodies, each with an area of about 17.7 sq m (Fridland, 1976b). The letters in the figure refer to map sheets of published soil surveys of counties in the United States: A, Ashtabula, Ohio; M, Montgomery, Illinois; H, Kaui Island, Hawaii; K, Kenosha, Wisconsin; C, Chickshaw, Mississippi; D-G, glaciated Dane, Wisconsin; S, Scotland, North Carolina; O, Okeechobee, Florida; D-D, unglaciated Dane, Wisconsin; and W, Ward, Texas. The distance between the disk- and the square-shaped data symbols for a particular terrain is a characteristic (called the "index of dominance of relatively large ESBs") of a soil cover pattern. Scaled values for the distances measured on the figure range from 100 for the Ashtabula County sample to 21 for the Ward County sample.

Some published soil maps provide extra information about the soil pattern by means of special symbols for ESBs that are less than 1 ha in size. The soil survey of Blue Earth County, Minnesota (Paulson et al, 1978) has map signs for "poorly drained soil with light colored subsurface horizon," "better drained soil," "wet spot," "high lime spot."

III. *Principal kinds of patterns* (as seen in plan view): Patterns may be thought of as relatively undifferentiated and differentiated. The CSB shown in Figure 4.5 is 78 sq km in area (Ashtabula County, Ohio) and is composed

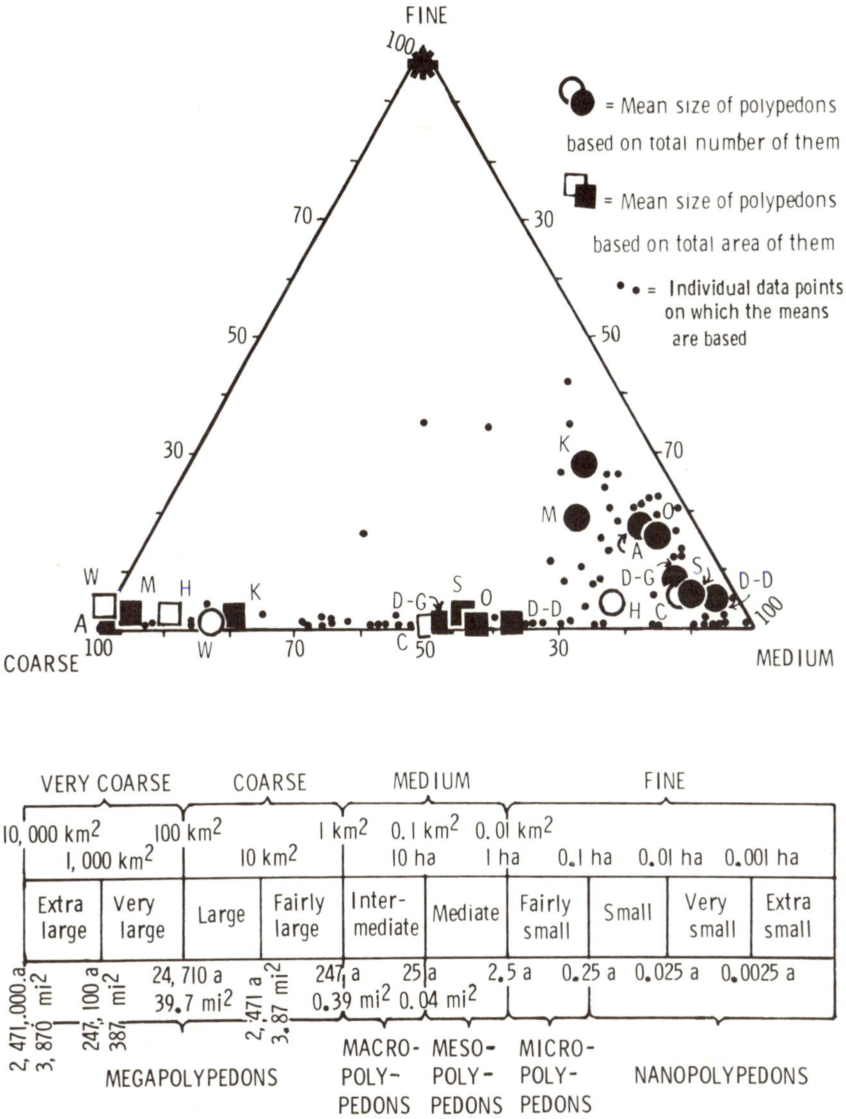

Figure 4.4. Guide for recording soil body size distribution in soil landscapes, with representative data plotted (see text).

largely of the poorly drained Sheffield silt loam (Typic Fragiaqualf, fine-silty, mixed mesic) (Reeder, Riemenschneider, and Reese, 1973). Figure 4.6 shows 4 quadrats each 1 sq km taken from this soil body. Quadrat A illustrates the undifferentiated condition. Of course, no undifferentiated terrain

Figure 4.5. Outline of a punctate combinational soil body (binary) in Ashtabula County, Ohio. (Inclusions not shown; see Figure 4.6.)

continues far. For example, quadrats B,C, and D are from the same CSB (Figure 4.5), and show progressive differentiation through interruption (5, 25, and 53% respectively) by slightly elevated bodies of a somewhat poorly drained soil, the Platea silt loam (Aeric Fragiaqualf, fine-silty mixed mesic). In a 24.2 sq km sample area of this CSB, 108 soil bodies (whole or partial) were counted (Habermann and Hole, 1980).

Differentiation with little or no alignment yields a patchwork pattern (Figure 4.7). A punctate pattern consists of ESBs arranged in a background that may consist of a large ESB (Figures 4.5, 4.6, and 4.8A), or of a patchwork (Figure 4.8B; inclusions are aligned). A line-centered pattern is depicted in Figure 3.16 and validated in Figure 4.9A. A point-centered pattern is shown in Figure 3.17. Intergrades and combinations of these principle soil landscape fabrics abound.

IV. *Origin of soil cover pattern*: Patterns that were set at "time-zero" by geologic agencies are numerous, as indicated in Figure 3.18. Many of these

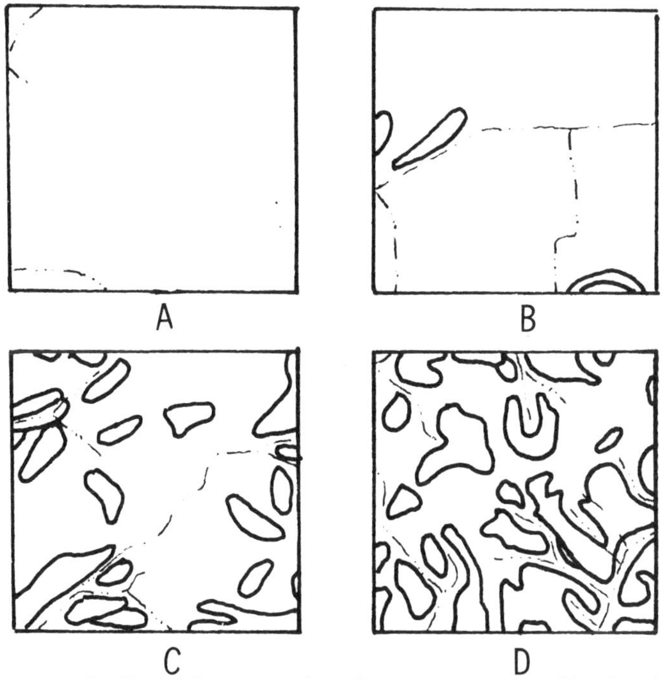

Figure 4.6 Four 1-km sq quadrats from a punctate combinational soil body (see Fig. 4.5), showing progressively (A–D) differentiated portions of it.

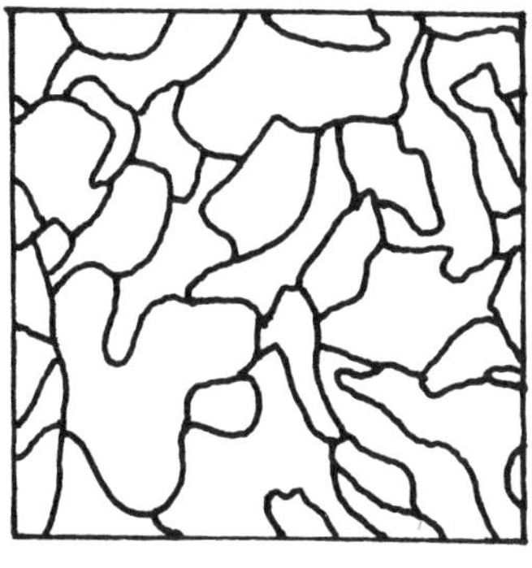

Figure 4.7. Patchwork pattern of soil cover in a 1 km sq quadrat from sheet 50 of the soil survey of Washtenaw County, Michigan. (Engel, 1977.)

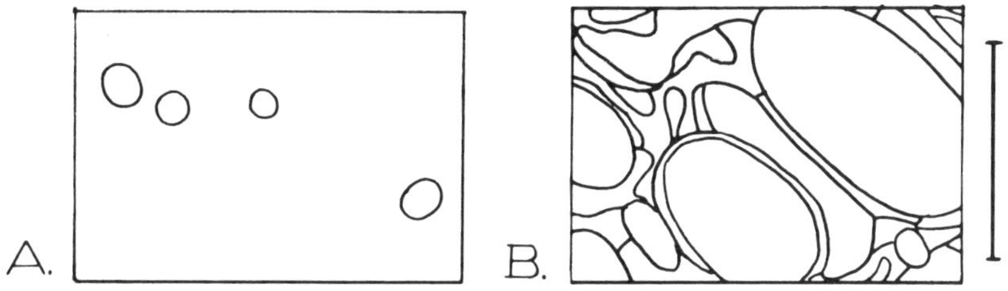

Figure 4.8 Two examples of punctate soil cover patterns. A. Elementary soil bodies of Pellusterts included in a large elementary soil body of Paleustoll in Floyd County, Texas. B. Large oval elementary soil bodies of Ochraquults in Carolina Bays set in a patchwork of irregularly shaped soil bodies of Paleudults in Scotland County, North Carolina. The bar represents 1 km.

patterns had original local relief features at a variety of scales, but some were relatively featureless. A featureless landform, illustrated at the right end of Figure 3.14 is the La Mesa surface, a mid-Pleistocene flood plain (elevation 1,320 m at present) of an ancient precursor of the Rio Grande River that now flows past Las Cruces, New Mexico at an elevation of 188 m. The figure illustrates the progressive destruction of the ancient surface and the associated extremely well developed desert soils (Petrocalcic Paleargids) (Gile and Grossman, 1979). Whatever the nature of the original land-

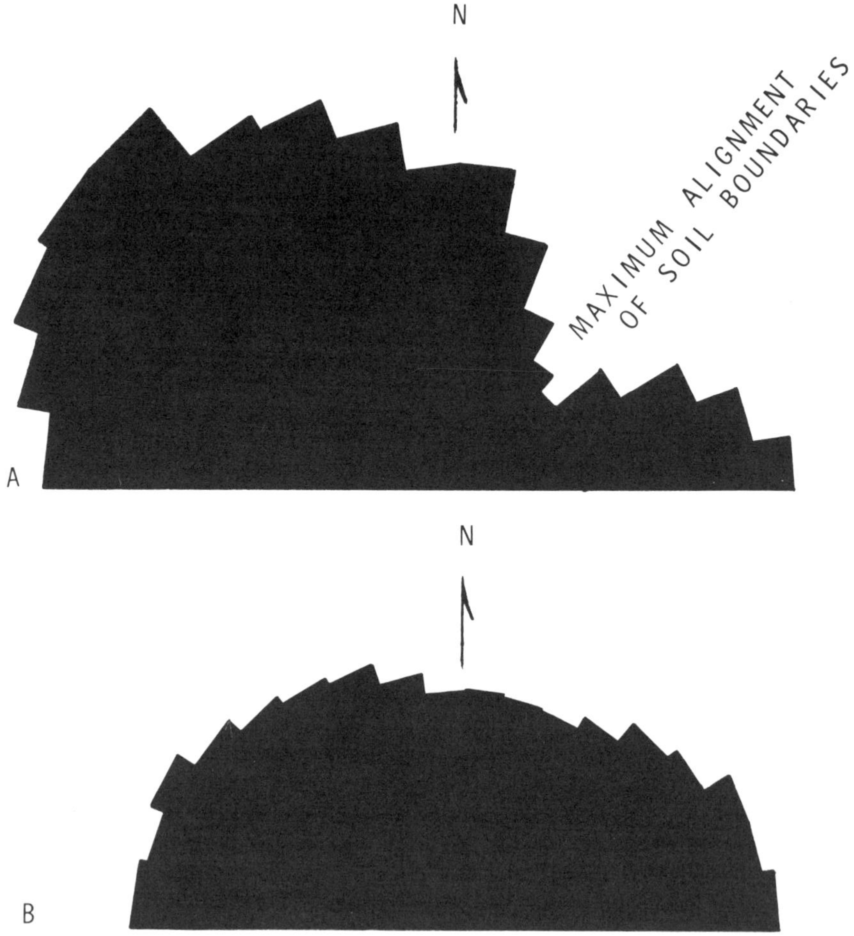

Figure 4.9. Compass half-roses reporting total number of soil boundaries intercepted by 20 slots/km (scale 1:10,000) in an overlay placed in 18 orientations at 10 degree intervals from due north to due south over a representative quadrat (1 km sq) of a drumlinoid ground moraine soilscape (above). Below is a similar diagram based upon a quadrat of a drumlin-free ground moraine soilscape in Dodge County, Wisconsin. (Fox and Lee, 1979.)

form, specific changes take place after time-zero that we may classify as geologic, pedologic, local-climatic, or biologic in origin. Dissection by stream erosion is primarily hydro-geologic, but the soil cover plays a role in determining how water moves in the landscape, as do local-climatic and biologic factors. Constructional events include accumulations of materials inside soil bodies and on their surfaces. In the old soils of the La Mesa surface, both clay and calcium carbonate accumulated to form impressive horizons of reddish sandy clay (B horizon) overlying a thick whitish rock-like "caliche" (K horizon; petrocalcic). At intervals, "pipes" developed in the K horizon which serve to conduct some storm waters vertically downward to underlying sand and gravel beds. Colluvial and local alluvial accumulations of soil materials are common on footslopes and toeslopes (Figures 3.11 and 3.12; Plate 4.1). Coalescing alluvial fans (at elevation 1,680 m sloping westward to 1,290 m) on the west side of the San Andres, San Augustin, and Organ mountains near Las Cruces constitute a piedmont that is 10 km wide and extends northward for tens of kilometers. Soil cover patterns that are degradational in origin are indicated in Figure 3.18 and Table 4.1. It is the task of the soil landscape analyst to identify and classify such patterns. Biotic overprints in soil cover are made by plants and communities of them along with associated fauna. Prairies yield bodies of soil with dark (mollic) surface soils that contrast with the domains of lighter colored A horizons developed under forest cover.

Plate 4.1 View of the Rio Grande valley as seen from a remnant of an alluvial fan originating from the Robledo Mountains west of Las Cruces, New Mexico, with flood plain and channel deposits (Torrifluvents) visible on the right. The uppermost surface visible in the left foreground is the late Pleistocene Pichacho surface described by Gile and Hawley (1972), eroded by Rio Grande drainage to form a distinctive arrangement of landscape units. (Photo by J. B. Campbell.)

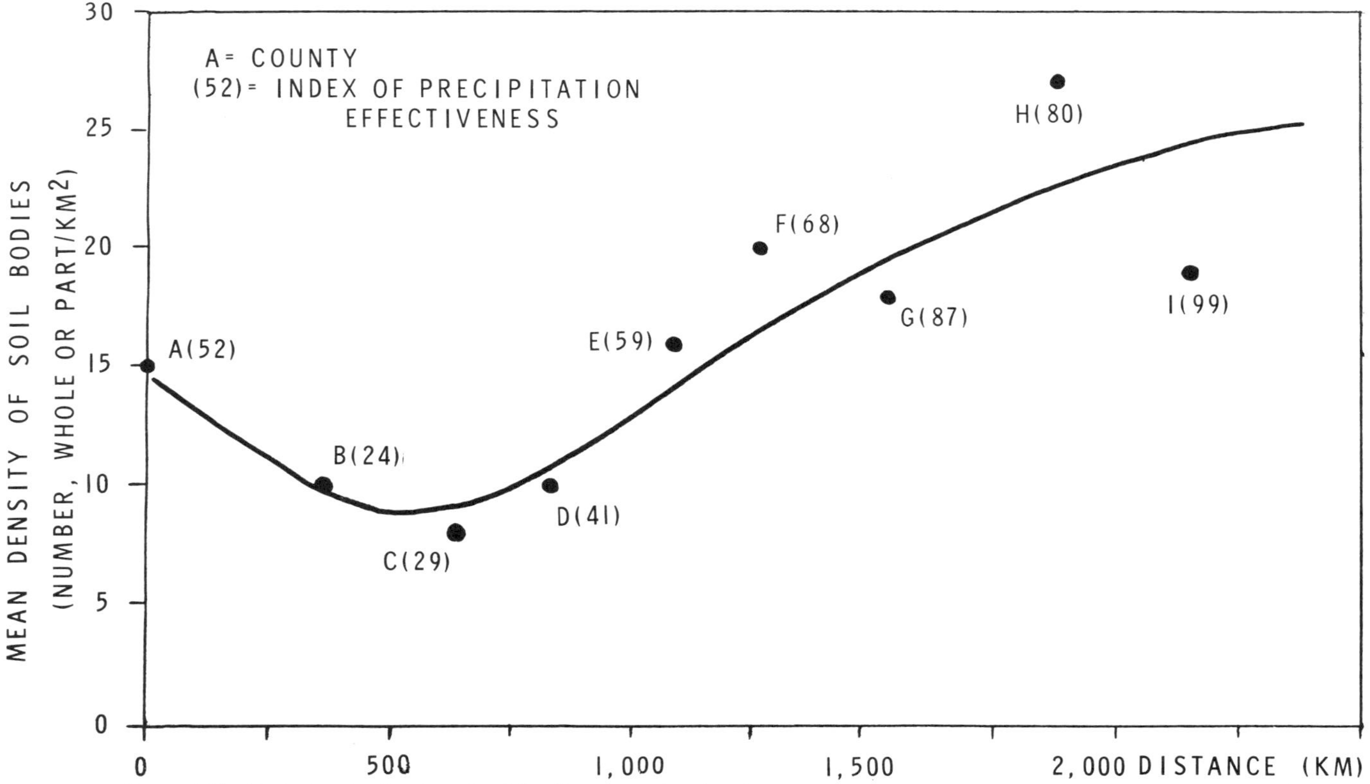

Figure 4.10. Variation in density of soil bodies per sq km in nine counties from Colorado to Indiana, approximately along the 11 degree (C) isotherm. Data are based on 32 randomly positioned quadrats in each county. Soil survey identification with date of publication: A = Delta-Montrose, Colorado (1967); B = Crowley, Colorado (1968); C = Lane, Kansas (1972); D = Rice, Kansas (1974); E = Shawnee, Kansas (1970); F = LaFayette, Missouri (1975); G = Dent, Missouri (1971); H = Logan, Illinois (1974); I = Davis, Indiana (1974).

V. *Population of soil bodies per unit area*: Hole reported a close correlation between mean density of soil bodies (numbers of them, whole or partial, per sq km; termed "intensity of soil map" by Laker, 1972), and mean soil boundary length (km per sq km) in a study of 288 randomly selected quadrats (each 1 sq km) in nine counties that lie approximately along the 11° (C) isotherm (annual) extending from the state of Colorado, to the state of Indiana. He also noted a correlation of population density of soil bodies per unit area with index of precipitation effectiveness (Thorthwaite, 1931) (Figure 4.10). The ratio between the mean count of soil boundaries per km of transect and the corresponding mean count of contour lines from a topographic map (at 3 m contour interval equivalent) was found to be 1.97 (8.53/4.33) on a Woodfordian glacial moraine (14,000 year-old surface) in northwestern Iowa, and 0.24 (10.21/42.54) on a Wisconsin loess-blanketed geomorphic surface with a Sangamon paleosol (> 70,000 years old) situated 40 km to the southwest. Corresponding measurements of relief per quadrat were 18.9 and 69.9 m, respectively, and of stream length, 1.53 km and 16.25 km, respectively (Hole, 1983).

The significance of the number and arrangements of nodes (junctions of soil boundaries, a species of points; Getis and Boots, 1978) per unit area (Figure 4.11) is yet to be determined.

A convenient device for collecting information from published soil maps is the template illustrated in Figure 4.12. In a sheet of cardboard one cuts a window the size of a desired quadrat (at the appropriate map scale), and stretches white thread to form north-south, east-west, and diagonal transects. Segments of these and additional segments along the borders of the window are darkened, as shown, in a manner that avoids intersection of transects. Proportionate pedologic content of a sample quadrat is determined by recording map symbols for points spaced at regular intervals along the transect segments and calculating percentage areas. Density of soil boundaries is determined from counts of intersections of transects with the boundaries. The information is reported as: (1) number of soil boundaries encountered per unit length, and (2) ratios between counts along transects lying normal to one another, to indicate degree of parallel orientation of boundaries. Hajek (1972) used a random transect method.

VI. *Composition*: Information concerning composition of soil cover may be given for quadrats (Figure 4.12), for individual combinational soil bodies (Figure 4.5), for watersheds (Figures 4.22 and 4.23), and other geographic units. The data curves for two soil landscape parameters in Figure 4.13 give signals that are less sensitive to a glacial boundary (position of an end moraine) than to soil landscape positions such as divides, valley walls, and valley floors. The soil mappers, however, used almost entirely different soil map legends on the two sides of the glacial boundary. Taxonomically, therefore, the boundary between the two complex combinational soil bodies is clear-cut. This example suggests that the true nature of soil cover patterns is less well understood than taxonomic treatment of them would lead us to believe. The combinational soil body of Figure 4.5 lies on a broad divide and is dynamically related by mono-flow to combinational soil bodies on surrounding hill-slopes. Figure 4.14 represents a quadrat in a complex com-

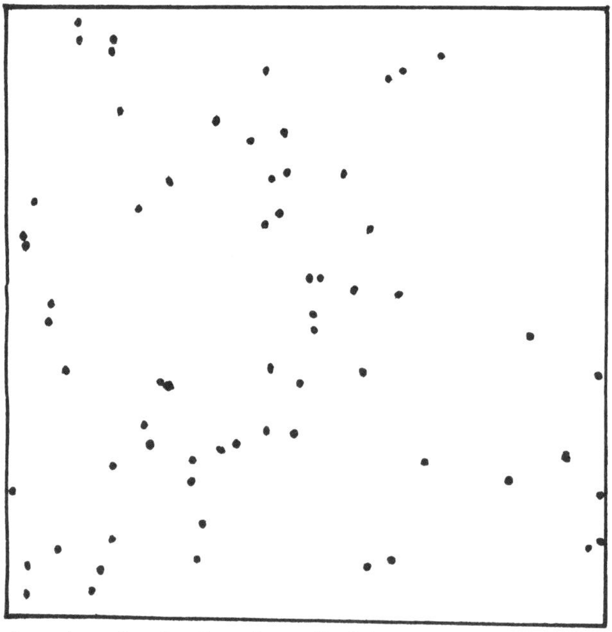

Figure 4.11. Location of nodes (junctions of soil boundaries in Section 11, T 89N, R 45 W in Woodbury County, Iowa. Total relief in this area is 53.1 M and stream length is 13.96 km.

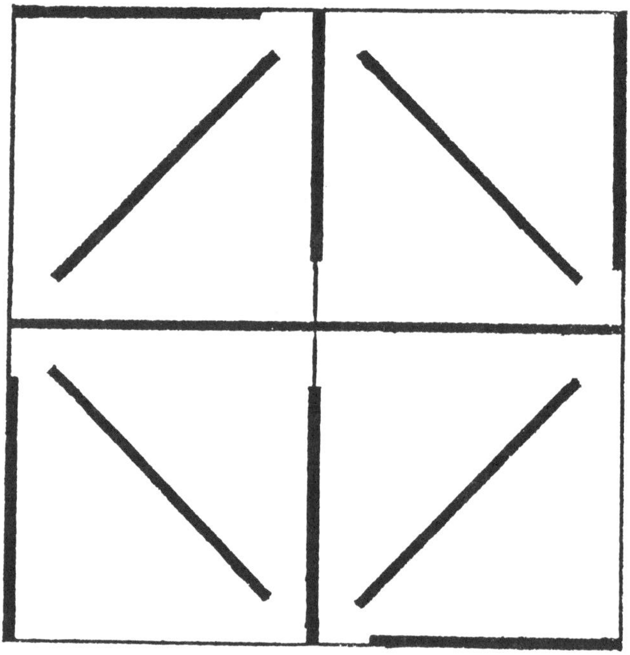

Figure 4.12. Diagram of a section (1 mi sq) showing with heavy lines the transects used to obtain data from soil maps and topographic quadrangles. The transects total 5.46 miles (8.79 km) in length.

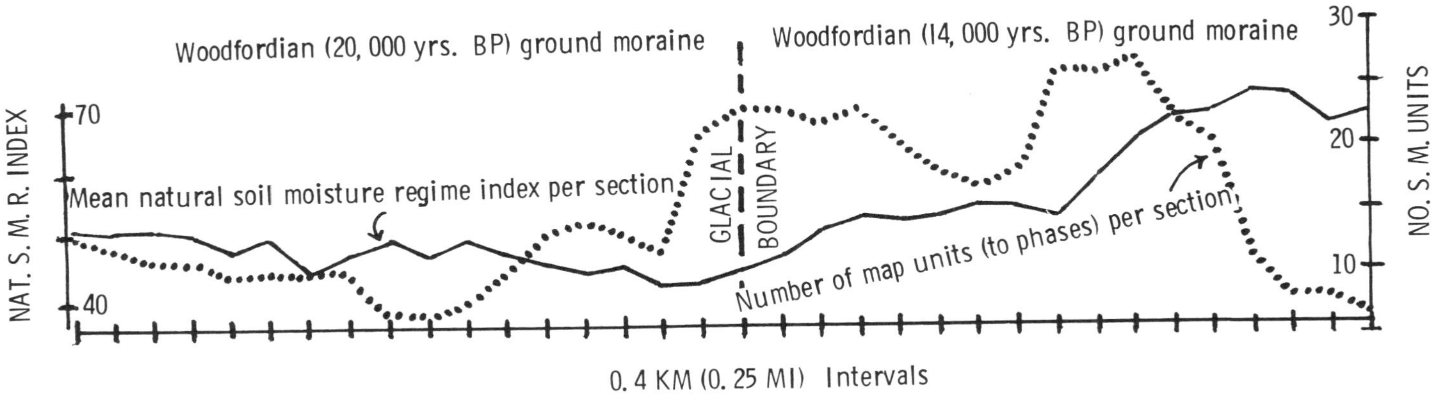

Figure 4.13. Data curves from an east-west transect across part of the soil map of Buena Vista County, Iowa. The curves were generated by shifting a window one-fourth mile in the same direction between data gatherings in terms of the two parameters represented by the two curves.

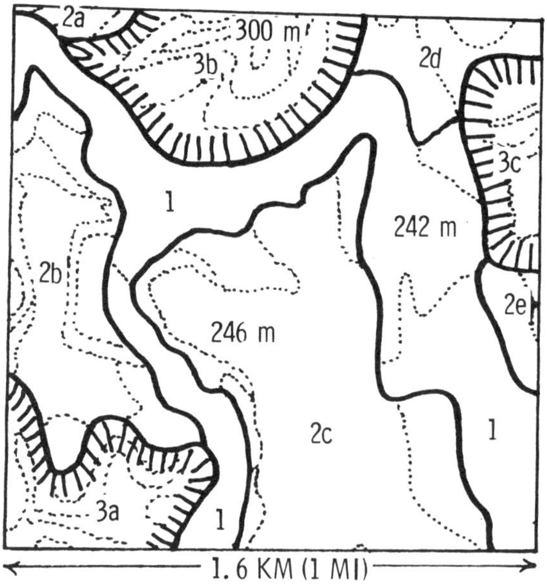

Figure 4.14 Combinational soil bodies delineated in three landscape positions (bounded by heavy lines): (1) Alluvial floodplain; (2) glacial outwash terraces; (3) Paleozoic bedrock hills discontinuously blanketed with loess (Section 26, soil map sheet 61 in Glocker and Patzner, 1978). Dotted boundaries delineate individual elementary soil bodies; representative elevations in meters above sea level are shown.

binational soil body in the the Driftless Area of Wisconsin, which is composed of portions of simple combinational soil bodies that are situated in three major landscape positions: 1a-1d, flood plain of Black Earth Creek; 2a-2d, loess-blanketed outwash terrace; 3a-3d, loess-blanketed Paleozoic rock-cored hills.

On a soil map of Military Ridge in Grant County, Wisconsin, Hole (1978) found a combinational soil body of 12,150 ha that contained only four soil series, as compared to thirteen soil series in a Carolina Bays combinational soil body in Scotland County, North Carolina. The mean numbers of soil series represented in a one sq km quadrat were 2.5 and 8.1, respectively, for the two areas. Habermann and Hole (1980) found only a limited relationship between relief and number of soil series present. In a sequence of terrains, showing increasing stream dissection of three geomorphic surfaces in northeastern Iowa, Hole (1983) noted in representative quadrats of published soil maps that the number of map units were, at the phase level, 12.5, 8.5, and 11.5 per sq mi, and at the series level, 11.6, 6.0, and 6.5. The trend is not a simple gradient.

Composition means proportionate extents of soil taxa. Generalized data for the chronosequences in northwestern Iowa, just mentioned, with relief per quadrat of 16.0 m, 19.7 m, and 47.1 m, respectively, are: (1) Haplaquolls on 59% of the area with islands of Hapludolls rising slightly above these wetlands, in the least dissected terrain; (2) Hapludolls occupying divides on 58% of the area, with Aquic Hapludolls and Cumulic Haplaquolls in dissecting drainageways in the second soil landscape; and (3), in the most dissected terrain, Typic Hapludolls occupying divides on 48% of the area, steep Udorthents on 36% of the area, and Cumulic Hapludolls and Udifluvents in drainageways in the remaining 16% of the area. The poorly drained soils occupy progressively less area and shift in landscape position from the divides to the drainageways. As Fridland (1976a) has noted, the geometric pattern of soil cover is commonly determined by geologic agencies. Habermann and Hole (1978) suggest that taxonomic content of soil cover pattern is influenced by zonal climate (compare a desert soilscape in Texas with a a humid tropical one in Hawaii and humid temperate one in the upper Mississippi Valley), by both climate and parent material together (note an Aquod-Aquent soil cover in Florida and a Vertisolic soilscape in Mississippi) and by relief (note differences between glaciated and unglaciated soilscapes, with 5 m and 13 m relief, respectively, in Wisconsin). Sandy initial material of the Florida terrain (Okeechobee County) have responded pedologically so sensitively to patterns of moisture regimes that there are nearly three times as many soil series per sq km there than in Ashtabula County, Ohio where Typic and Aeric Fragiudalfs developed in fine textured materials (Figure 4.5).

VII. *Soil cover diversity*: Soil cover diversity has to do with density of soil bodies and variety of soil species present.

Taxonomic contrast. We have developed the following expression for estimating taxonomic contrast ("degree of dissimilarity" according to Dan J. Yaalon, pers. communication, 1978):

$$\frac{E_{na-9}}{E_{np}} \cdot \frac{U_{ne}}{1} = \text{Index of taxonomic contrast}$$

E_{na} is the actual number of taxonomic elements present in the map legend that contains five map units (soil species) for two combinational soil bodies in a given terrain. E_{np} is the potential number of taxonomic elements if the terrain displayed its maximum possible contrast. U_{ne} is the effective number of units in the map legend. Table 4.2 presents a working model for applying the formula. The model depicts two combinational soil bodies (CSB#I and CSB#II). The taxonomic elements are numbered 1 through 9. The row sums for E_{na} reports the number of kinds of elements present in each column. The row of values for E_{np} reports the greatest contrast possible for the terrain in question. We may note that a large elementary soil body composed entirely of the first soil species in the list would show no taxonomic contrast, even though 9 elements were present. Therefore, the formula states E_{na-9}. The ratio of $(28-9)/(36)$ applies to both CSBs. The effective number of soil map units takes into account proportionate extents. The pro-

Table 4.2 **Examples of Calculation of Taxonomic Contrast of Two Quadrats of as Many Combinational Soil Bodies.**

Five soil map units

	1 Slope	2 Erosion	3 Texture	4 Mineralogy	5 Soil temperature regime	6 Subgroup	7 Grt. Group	8 Suborder	9 Order	Proportionate Extents in Quadrats of CSB #I %	CSB #II %
	B	2	fine-silty	mixed	mesic	Typic		Argiudoll		20	90
	C	3	coarse-loamy	mixed	mesic	Entic		Haplaquept		20	6
	D	3	fine-loamy	mixed	mesic	Mollic		Fragiudalf		20	1
	B	1	fine	mixed	mesic	Typic		Udifluvent		20	1
	A	1	fine-loamy	mixed	mesic	Cumulic		Hapludoll		20	1
E_{na}[a]	4 + 3 + 4 + 1 + 1 + 4 + 4 + 3 + 4 = 28									5	1.099
E_{np}[b]	5 + 3 + 5 + 2 + 1 + 5 + 5 + 5 + 5 = 36									5	1.099

$$\frac{28-9}{36} \times \frac{5}{1} = 2.635 \qquad \frac{28-9}{36} \times \frac{1.099}{1} = 0.579$$

Note: [a]Actual number of taxonomic elements.
[b]Potential number of taxonomic elements in the soil landscape under study.

cedure is to divide each percentage in a given column in a right-hand side of the table by the largest percentage, and then to add the values obtained. The sums are the effective numbers of units, which in these two cases are 5 and 1.099, respectively. The products of the two items are the indices of contrast: 2.635 for #I and 0.579 for #II.

Calculation of contrast may be based, not on taxonomy, but on actual soil properties, such as average pH and content of clay, silt, or organic matter to a depth of 1 meter. Van Heesen (1970) transformed a taxonomic soil map of a 162 ha parcel into a water-table-class soil map. Westin (1976) found that heterogeneity of results of soil tests in South Dakota correlated regionally with climate and vegetation, but locally with parent material, slope, and age, which provide bases for contrast between adjacent soil bodies.

Soil Body Size Distribution. Figure 4.4 shows a triangle on which representative data are plotted from ten published soil map sheets (Habermann and Hole, 1980). Sizes of whole soil bodies may be determined by cutting the map up, and weighing the parts within an interval of time short enough to avoid appreciable weight change of the paper due to gain or loss of moisture from the air, then by comparing weights of the parts to weights of paper squares representing standard areas. Counting dots arranged in a grid on otherwise transparent overlay sheets is a method of approximating areas of soil body delineations. Planimetric and other procedures may also be used. Relationships between size of background body and included bodies are of interest. Conceivably the trend in Figure 4.6 B-C-D might be continued to the point of forming a tight cluster of elementary soil bodies with only a thin intervening strip remaining of the background body.

Number of Soil Bodies Per Unit Area. Studies by Habermann and Hole (1980) suggest that counts of numbers of soil bodies, whole and partial, per quadrat gives numbers that are 1.2 to 2.2 times larger than the mean number of whole bodies per quadrat for a large area. Soil maps may be generalized in a variety of ways (see Chapter 8), and numbers of soil bodies per unit area (as represented on a map) change accordingly.

Index of Heterogeneity. A simple index of heterogeneity is the product of the number of soil bodies (whole or partial) per unit area (the density of soil bodies), and the number of soil mapping units in the legend for the unit area. A more accurate index of heterogeneity is the product of the index of taxonomic contrast (see above) and the density of soil bodies. Two representative soil body density values reported by Habermann and Hole are 7 and 25 per sq. km. If these are multiplied by the two indices of contrast given in Table 4.2, the resulting indices of heterogenity range from 18.45 to 65.88 for CSB#I and from 4.05 to 14.48 for CSB#II.

Number of Soil Map Units Per Unit Area. Figure 4.13 shows that the number of taxonomic units (to the soil slope and erosion phase level) ranged from 6

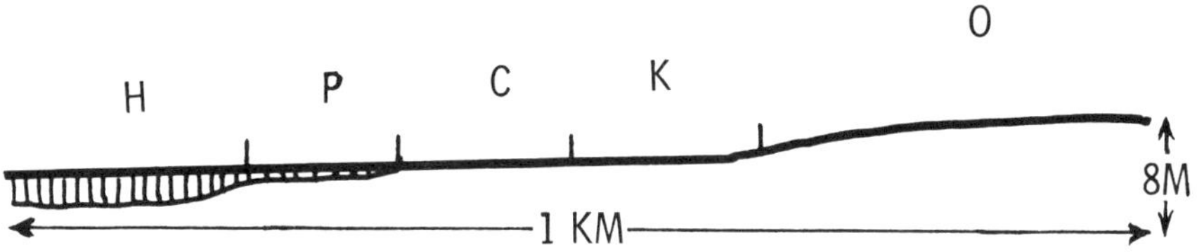

Figure 4.15. Generalized cross-section through a soilscape segment in the northwest corner of Green Lake County, Wisconsin, U.S., illustrating arrangement of principal (end member: H and O) and transitional (P, C, and K) components. H = Houghton Muck (Typic Medisparist); P = Palms Muck (Terric Medisaprist); C = Colwood Silt Loam (Typic Haplaquoll); K = Kibbie Silt Loam (Aquollic Hapludalf); O = Oshtemo loamy fine sand (Typic Hapudalf). (Anderson and Gundlach, 1977).

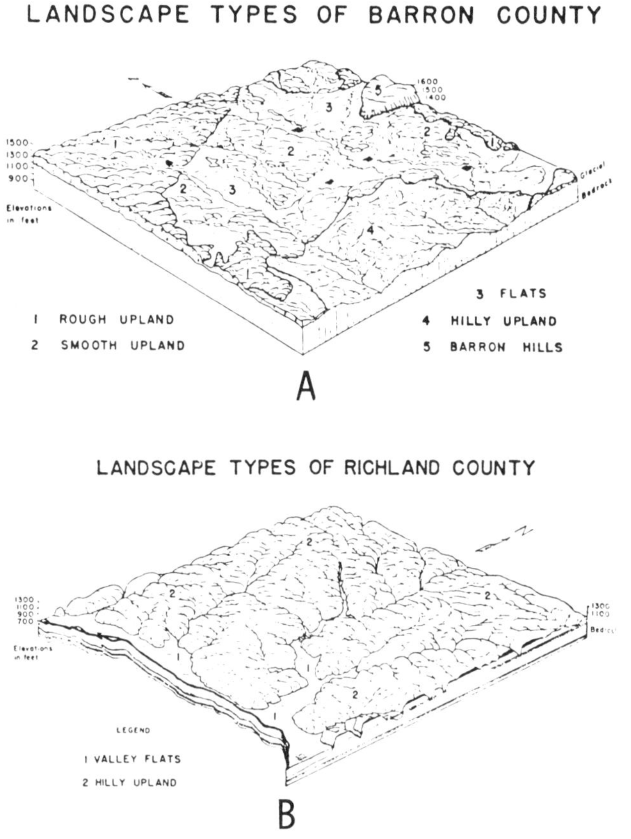

Figure 4.16. Block diagrams of two counties in Wisconsin, showing major combinational soil bodies in each (Hole, 1953).

to nearly 30 per 1 sq mi quadrat of a soil map in northwestern Iowa (Hole, 1983). Variations in this parameter from quadrat to quadrat within a given published survey reflect variations in number of contrasting soil landscape positions and patterns included within the quadrats.

Number of Soil Landscape Positions. Five soil landscape positions are shown in the cross-section of a catenal surface having 8 meters of relief (Figure 4.15). The end members may be called *principal components* and the intervening ones *transitional components.* In a terrain with local relief of about 60 m (Figure 4.16B), Hole noted (on the basis of 50 quadrats) a minimum of five landscape positions, which are shown for bodies of the Dubuque soil series (Typic Hapludalfs developed in thin loess over residual clay on dolomite) and associated soils down-slope (Figure 4.17). The wet soil (Aquent) below the well-drained upland soils (bounded by T-T) is in the soil landscape position 5.1, which means that at 90% of the sites it lay in the fifth position and at 10%, the sixth position ([90 × 5] + [10 × 6] × 0.01). The count of soil landscape positions is the minimal count of soil bodies down the slope. An alternative procedure would be to make body counts along many transects and report the mean value. This geographic analysis reports the soil landscape habits of individual soil series. The Dubuque soil series has a soil landscape position of 1.3 ([70% × 1] + [30% × 2 × 0.01]). At 30% of the sites, the Fayette silt loam, a Typic Hapludalf formed in deeper loess, occupied the first position, displacing the Dubuque soil bodies to second position.

In glaciated terrain (Figure 4.16A and 4.18), with local relief half of that of the area just discussed, and with less steep slopes, bodies of the well-drained Otterholt soil always occupied the first position and in such instances (at 40% of the 50 sites studied) bodies of the moderately well drained Spencer soil lay in second position. In the absence of the Otterholt soil (on 60% of the hill crests) the Spencer soil assumed the first position. In this smooth upland, Spencer soil bodies had a soil landscape position index of 1.4 and wet soils (Haplaquolls) were at 3.6, in the presence of Otterholt bodies; and at 2.9 where the Spencer was in first position. Analysis of both stream-line and valley wall catenas of soils has been outlined by Hole. (See Appendix II, p. 148.)

Soil Moisture Regime Diversity. Hole (1978) devised a soil moisture regime index, SMRI (sometimes called the natural soil drainage index), based upon arbitrary assignment of 0 to a monolithic granite outcrop containing no water, 100 to a saturated deep peat, and intermediate numbers to mineral soils that are well drained ("zonal"): (40), moderately well drained (50), somewhat poorly well drained (60), poorly drained (80), and very poorly drained (90). (See Appendix II, p. 149.) Ratings between 0 and 40 are assigned to soil bodies with little moisture storage capacity. A soilscape consisting entirely of saturated peat bodies would have a SMRI of 100. A soilscape composed entirely of bodies of "zonal" soils such as Chernozems

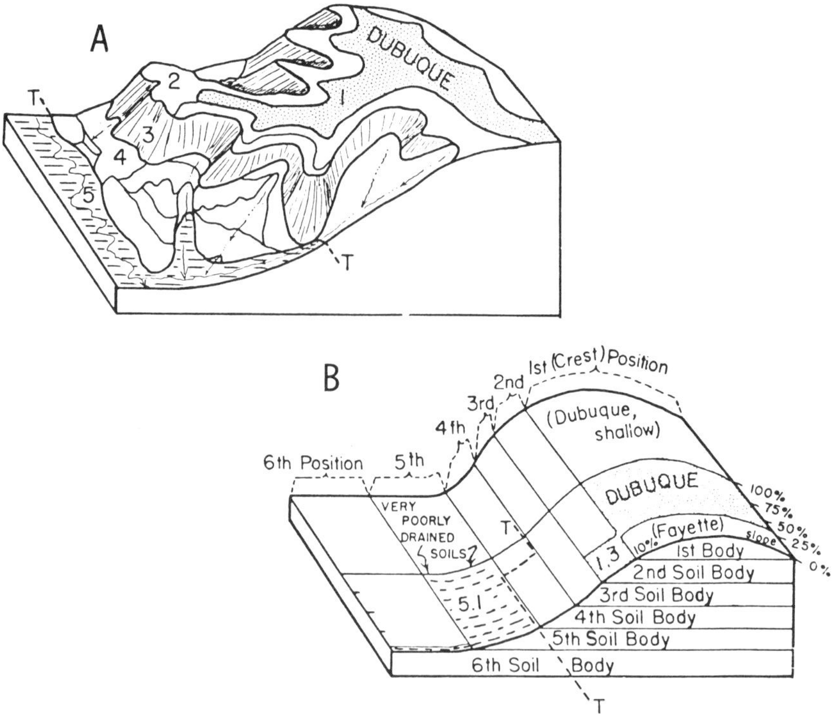

Figure 4.17. Block diagrams of representative portions of the hilly upland of rich land county (see Fig. 4.16B): A. Quadrat showing a Dubuque soil body in first position, with a minimum count of five soil bodies down-slope across well-drained soils (bounded below by line T-T) to a poorly drained soil body (Aquent). B. Statistical diagram, representing 50 1 mi sq quadrats, shows that the Dubuque soil lies in first position in 70 percent of the quadrats and in second position in 30 percent.

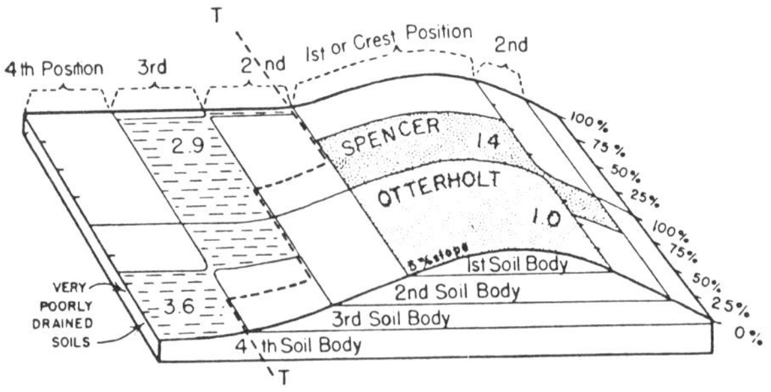

Figure 4.18 Statistical block diagrams representing 50 1 mi sq quadrats in the smooth upland of Barron County (see Figure 4.16). The Otterholt soil occupied first position whenever present; the Spencer soil occupied first position in 60 percent of the instances and second position in 40 percent.

(Borolls), would rate 40. The range of regimes for a given combinational soil body might be from 20 to 100 and average 60 (by proportionate extent); or from 40 to 80 and also average 60. Both parameters are needed to characterize a terrain, as well as number of regimes present. Habermann and Hole (1980) reported from ten different soil maps SMRIs that ranged from 38 for an Aridisol landscape in Texas, to 77 for an Aquod-Aquent landscape in Florida. In order to place the indices on a global scale, we suggest multiplying the SMRIs just described by generalized climatic ratings derived from discharge (runoff) maps of Baumgartner and Reichel (1975). Table 4.3 presents proposed numerical moisture regimes. Soil bodies of comparable texture (such as loam) that are naturally well drained, with local SMRI of 40, have global moisture regime indices of 0.4 (40×0.01) in a desert setting, 4 on a steppe, 120 in a steppe-prairie transition, 1,000 in a prairie (such as Illinois), 2,000 in a boreal forest, and 3,200 in a maritime per-humid setting (as in Laborador). Very poorly drained mineral soils, such as Aquolls, with local SMRI of 90, have in these same landscapes global moisture regime indices of 0.9, 9, 270, 2,250, 4,500, and 7,200, respectively. Global equivalents for the local SMRIs 38 and 77 mentioned above are 0.38 and 1,540. These arbitrary ratings may serve to stimulate refinement of indices of local and continental diversity of soil cover with respect to soil moisture regime.

Soil Body Shape Index. This relates the shape of a soil body to that of a circle by means of the ratio of the length of the soil boundary to the perimiter of a circle of the same area. Juday (1914), in Wisconsin, used a similar ratio to determine "shore development" indices of lakes, which he observed became more circular with increasing age. The opposite trend is common for soil bodies, which tend to become more irregular in shape with age as a

Table 4.3 Generalized Categories of Moisture Regimes

Climatic moisture regime		Mean annual runoff index[a]	Soil moisture regime[b]
A. Perhumid		> 70	Perudic
B. Humid		70-60 60-50 50-40 40-15	Udic
C. Subhumid	Moist Dry	15-10 10-5	Ustic
D. Steppe		5-0.1	Xeric or Ustic-Aridic
E. Desert		0.1-0.01	Aridic, Torric

Source: [a]Generalized from world runoff maps by Baumgartner and Reichel (1975).
[b]Adapted from Soil Survey Staff (1975).

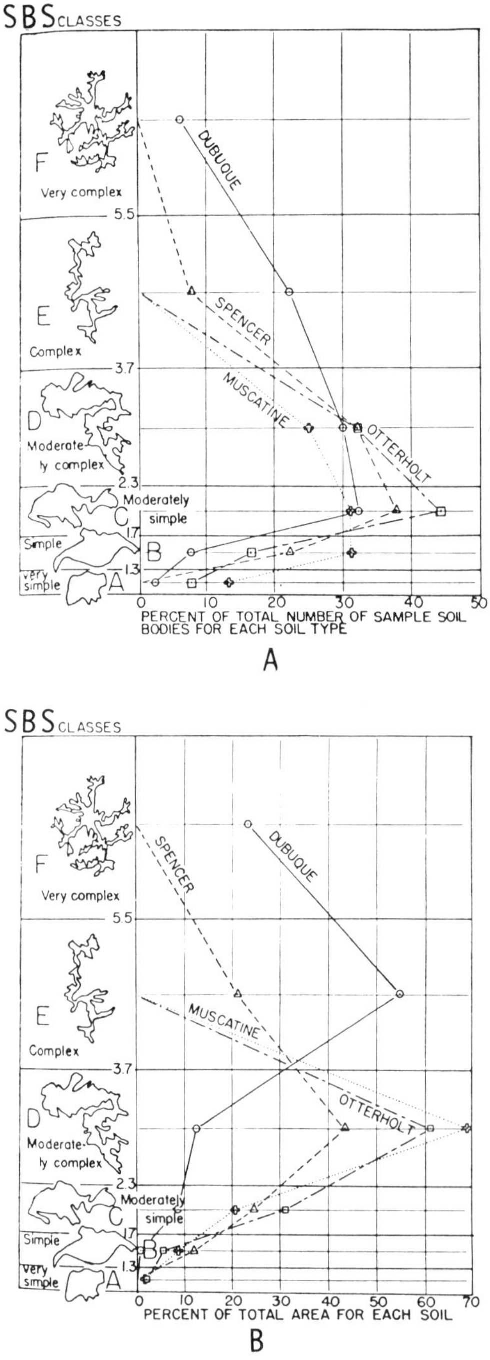

Figure 4.19. Soil body shape distribution of sample bodies representing four soil series on published soil maps (Hole, 1953).

result of geologic stream dissection. In fact Fridland (1976a) uses the term "index of dissection" for this ratio. In a study of two terrains, (1) the smooth upland of glaciated portions of Barron County, Wisconsin, with local relief of 30 m (Figure 4.16A), and (2) the upland of unglaciated Richland County, Wisconsin, with local relief of 60 to 90 m (Figure 4.16B), Hole (1953) found that bodies of the Otterholt silt loam (well drained) in the smooth upland are less dissected than are bodies of the moderately well drained Spencer soil (Figures 4.18 and 4.19), which lie closer to drainageways. In the unglaciated upland, well drained Dubuque soil bodies show a wide range of shapes, including some complex and very complex bodies that are very large (Figure 4.19). Irregularity of large soil bodies is depicted more faithfully on soil maps than is that of small soil bodies, whose shapes are commonly smoothed in the preparation of a soil map. Muscatine soil bodies (moderately to somewhat poorly drained) lie on flat ridge crests, remote from drainageways, and are less dissected than are Dubuque soil bodies.

Hole (1978) related the shape indices just discussed to three major classes of shapes (Figure 4.20). Dan H. Yaalon (pers communication, 1978) suggests "equant" and "ellipsoid" as synonyms of "disk" and "spot", respectively.

Bunge (1962) (see p. 32 of Fridland, 1976a; Poore and Huddleston, 1983) proposed a quantification of shapes of delineations by enclosing them in polygons (see body number 1 in Figure 4.21), then measuring distances between alternate apices in a counter-clockwise direction; and summing the distance and the squares of the distances, and by certain ratios (Table 4.4). Marked irregularities in soil boundaries affect the results of the manipulation, as indicated.

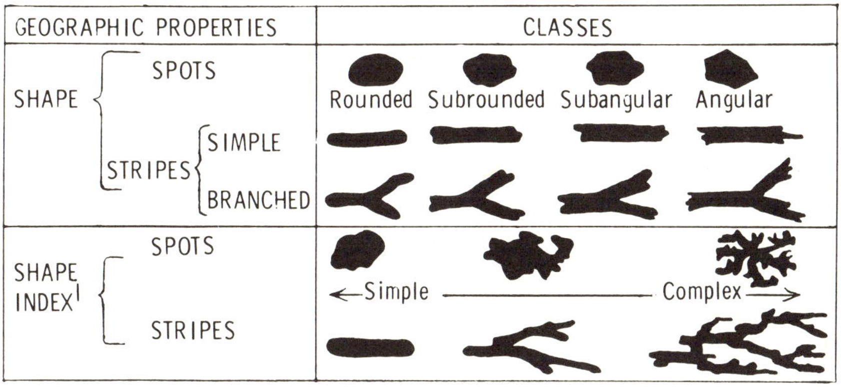

Figure 4.20. Classes of shapes of soil bodies (Hole, 1953).

Degree of Distinctness of Soil Boundaries. It is rare indeed to find a soil map that reports soil boundary widths (Table 4.5), although some of the large scale soil maps of the Soil Survey of England and Wales have used dashed lines to symbolize indistinct boundaries. If the soil map analyst is provided

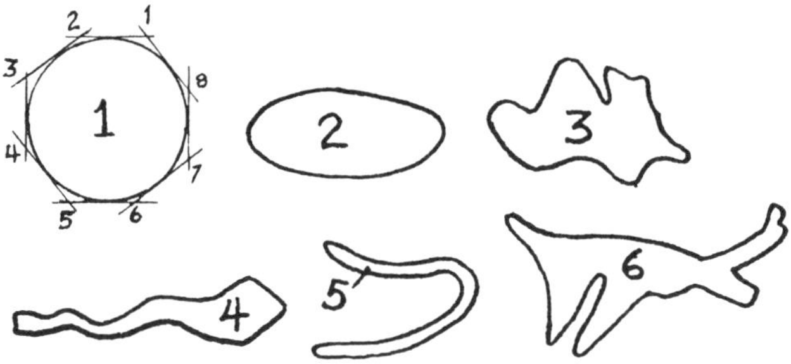

Figure 4.21. A circle and five delineations of soil bodies taken from sheet 5 of the soil survey of Green Lake, Wisconsin. (Anderson and Gundlach, 1977).

Table 4.4 Characteristics of Shapes of Six Soil Bodies Shown on Sheet 5 of the Soil Survey of Green Lake County, WI (See Figure 4.15)

Body number	Sum of distances (mm) between apices[a] (number of distances measured)	Sum of squares of distances between apices[a]	Ratio of sum of squares to sum of distances	Ratio between maximum length and maximum width of body	Name of shape[b]
1	156 (8)	3,042	19.5	1.00	Disk
2	155 (8)	3,181	20.5	2.36	Spot
3	152 (16)	1,644	10.8	2.10	Spot
4	199 (17)	2,870	14.4	5.40	Stripe
5	215 (15)	3,591	16.7	19.00	Stripe
6	285 (24)	3,851	13.5	2.40	Spot

Note: [a]Soil body number 1 (Figure 4.21) is enclosed in a polygon with 8 apices. Distances were measured between apex no. 1 and apex no. 3, between apex no. 2 and apex no. 4, and so on.

[b]Defined by the ratio in the preceding column: 0-2, disk; 2-5, spot; > 5, stripe.

Table 4.5 Suggested Classes of Soil Boundary Widths

	Class	Definition
Discrete	Very sharp	< 0.3 m
	Sharp	0.3-3 m
	Distinct	3-5 m
	Gradual	5-10 m
Continual	Diffuse	10-25 m
	Transitional[a]	> 25 m

Source: Adapted from Fridland (1976a).

Note: [a]On highly detailed soil maps this boundary may be converted into a transitional soil body.

with information on width of soil boundaries and degree of taxonomic or pedologic contrast between neighboring soil units, he or she may proceed to develop a set of indices of boundary distinctness suited to a particular terrain.

Taxonomic Rank of Soil Boundaries. Although soil boundaries on maps have the same weight of line in most instances, different segments of the boundaries (between nodes) may serve at different combinations of taxonomic ranks. Hole (1978) used 9 ranks that differed somewhat from the sequence shown in Table 4.2. His ranks, by soil categories (Soil Survey Staff, 1975a) were assigned as follows: 1) phase; 2) type; 3) variant; 4) series; 5) family; 6) subgroup; 7) great group; 8) suborder; 9) order. He measured length (km) of soil boundary segments per quadrat that served at the 9 levels. Some soil boundaries serve only at the first level; some serve at many levels simultaneously. Table 4.6 summarizes results of a study of soil boundary diversity in two watersheds in Wisconsin (Figures 4.22 and 4.23). Soil contrast is reported as (1) the proportionate lengths of soil boundaries according to maximum categorical levels at which they serve (columns 3 and 4 in the table); and (2) the proportionate lengths of soil boundaries of different complexities (average number of ranks at which a boundary of a given minimum rank serves simultaneously). In the quadrat from the Driftless Area (Figure 4.22) 97% of the area is occupied by Mollisols, and the soil boundaries have an average maximum soil rank of 3.8 as compared with 6.6 in the quadrat of the glaciated terrain (Figure 4.23) where Mollisols occupy 49% of the area. The average number of soil series per sq km was found to be 9.8 in the glaciated terrain (based upon eight quadrats) and 4.1 in the

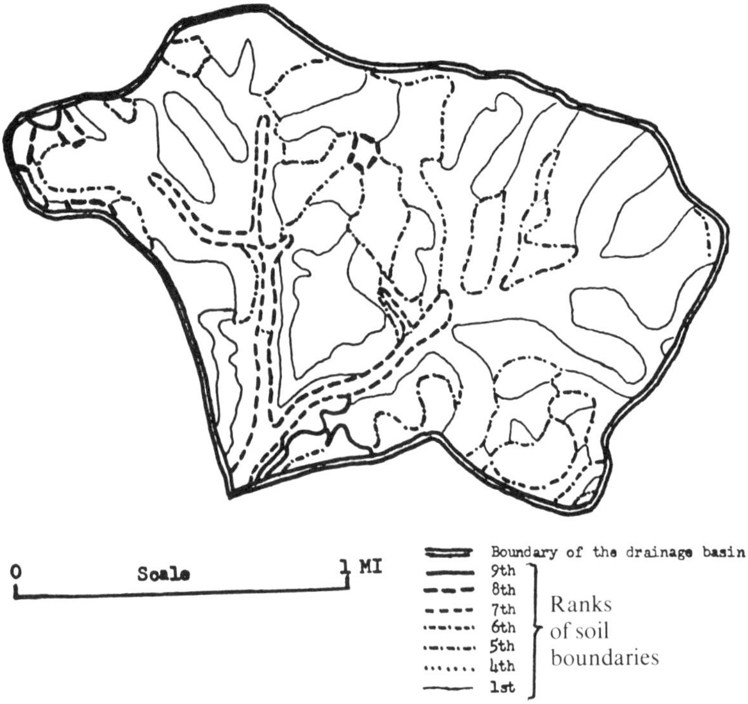

Figure 4.22. Soil boundaries expressing degree of contrast between adjacent soil bodies in a fourth order drainage basin (3.06 sq km) situated 2 km northwest of Dodgeville, Iowa County, Wisconsin, in the "driftless area." Relief is about 72 m. Scale is 1:20,000. (Klingelhoets, 1962.)

Driftless Area (based upon six quadrats). In the drainage basin in the glaciated area 5 soil orders were represented; 3 in the basin in the Driftless Area. Chapter 8 presents further discussion of the cartographic symbolization of taxonomic rank.

Degree of Chronological Uniformity. If we accept the definition of Entisols as very young compared to many mineral soils of other orders, then the relative areas occupied in a terrain by Entisols and older soils are indices of degree of chronological uniformity and diversity. In the quadrat of drumlinoid ground moraine (with 15 m of local relief; Figure 3.16) Entisols (Fluvents) occupy 2.4% of the area and Alfisols (65.9%), Mollisols (31.7%), and Histosols (1.6%) account for the rest. Hapludolls occupy 26% of the area and may be considered to be chronologically intermediate between Entisols and the Alfisols and Mollisols that have argillic horizons (71.6% by area). A quadrat (Sec. 26, T. 87 N., R. 45 W., with 41 meters of relief) in Woodbury County, Iowa, of a dissected terrain in which a Sangamon soil is present under loess cover, has 28% Entisols by area (57% Orthents, 5% Fluvents), and 28% Mollisols (all without argillic horizons).

Figure 4.24 for Bartholomew County, Indiana, shows relationships between geomorphic surfaces of different ages and normalized soil profile

Table 4.6 A Comparison of Diversity of Soils in Two Fourth-order Watersheds in Wisconsin[a]

		Proportionate extent[d]		Average degree of complexity of soil boundaries[e]	
Soil boundary rank number[b]	Example of category[c]	Watershed in the "Driftless Area" (%)	Watershed in the glaciated area (%)	Watershed in the "Driftless Area"	Watershed in the glaciated area
9	Mollisol	2 } 25	57 } 73	5.6	5.3
8	Udoll	3	5	5.8	3.3
7	Argiudoll	16	trace	4.5	—
6	Typic	4	11	3.6	3.5
5	Fine-silty, over clayey, mixed, mesic	31	2	2.6	2.5
4	Dodgeville	trace	1	—	1.9
3	Deep phase	0	0	—	—
2	Silt loam	0	trace	—	—
1	6–12% slope	44	24	1.0	1.0
Average		3.8	6.6[f]	2.4	3.9

Notes: [a] A watershed 2 km N.W. of Dodgeville, Wisconsin in the "Driftless Area" (Fig. 4.22), and a watershed 2 km N.W. of Berlin, Wisconsin in Woodfordian glacial drift terrain (Fig. 4.23).

[b] Soil boundaries are ranked here according to the highest level of classification of the soil bodies separated: 1 = soil phase; 2 = soil type; 3 = soil variant; 4 = soil series; 5 = soil family; 6 = soil subgroup; 7 = soil great group; 8 = soil suborder; 9 = soil order.

[c] See Soil Survey Staff, 1975a.

[d] By total measure with a map measurer. Classification of 50 soil boundaries intersected by transects gave results within 10% of the total measurement. The average = % length × rank number, summed × 0.01.

[e] The complexity index is the average number of levels (1 through 9) at which a soil boundary serves simultaneously. Proportionate lengths were based on measurements made with a map measurer, and indices of complexity were multiplied by proportionate lengths to yield average indices. Results obtained by transects intersecting 50 soil boundaries came within 10% of the results given above, based on measurement of total lengths of soil boundaries. The average = complexity index × % of length, summed × 0.01.

[f] If the sequence of rank numbers is reversed, these two average ratings come to 6.2 and 3.5, respectively.

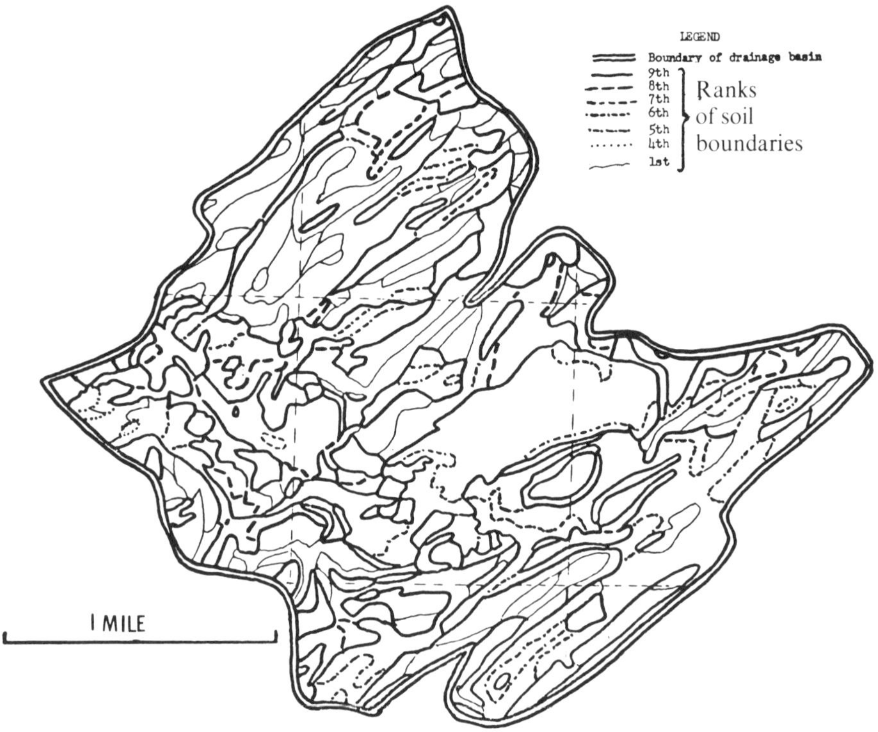

Figure 4.23. Soil boundaries expressing degree of contrast between adjacent soil bodies in a fourth order drainage basin (9.8 sq km) situated 2 km northwest of New Berlin, Waukesha County, Wisconsin, on Woodfordian Drumlinoid Moraine. Relief is about 73 m. Map scale was 1:15,840. (Steingraeber and Reynolds, 1971.)

development indices (Bilzi and Ciolkosz, 1977; Harden, 1982) for 4 soil profiles. If the curve is credible, then the Cincinnati soil profile on valley wall slopes is close to 45,000 years old.

The 400 sq mi (1,036 sq km) study area of the Desert Soils Project (Gile and Grossman, 1979) has great chronological diversity. Soils range in age from current Fluvents of the Rio Grande flood plain to Petrocalcic Paleargids of Mid-Peistocene age (700,000 B.P.).

Degree of Openness. Topographic, pedogeomorphic and hydrologic information is needed to characterize soil patterns with respect to degree of closedness (surface drainage is internal; Figures 3.10 and 3.17; is semi-closed, Figure 3.13) and of openness (drainage is external; Figure 4.25). Ruhe (1969) has thoroughly discussed this topic.

Length of Drainageway Per Unit Area. In a study of 4 geomorphic surfaces in northwestern Iowa (Hole, 1983), total length of stream drainageways (indi-

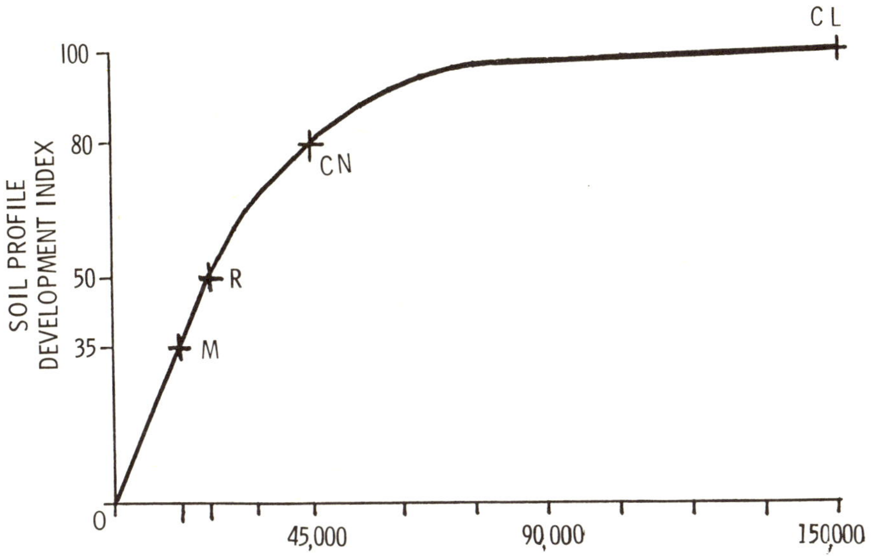

Figure 4.24 Relationship between ages of several geomorphic surfaces in central Indiana, and soil development indices (normalized) for representative pedons of four soils developed in glacial till with loess cover that is thin in the Miami and thick in the other three series. Miami (M, 14,000 years old), Russell (R, 20,000 years old), and Clermont (CL, 15,000 years old) soil are on divides on Woodforidan (M, R) and Illinoian (CL) glacial till plains. Cincinnati (CN) soils are in walls of valleys that dissect the Illinoian till plain.

cated by both intermittent and perennial stream symbols on topographic and soil maps) was measured per quadrat (1 sq mi). This distance varied from 0.64 on a young moraine to 13 km in a dissected older terrain. The relevant normalized data curve is labeled number "1" in Figure 4.26. Stream length can be determined with the aid of a map measurer on published soil and topographic maps. (In making such measurements, landscape analysts should be aware of the numerous inaccuracies that have been reported in comparing actual stream length with representations on the usual topographic maps (Chorley and Dale, 1972; Mueller, 1979).) A plot of data on which curves numbered 11 and 12 were based for Figure 4.26 were plotted in Figure 4.27, with the result that data points were segregated into clusters that represented 4 geomorphic surfaces.

Pedologic Sequences Along Drainageways and Normal to Them. Figure 4.28 is a plan diagram for four kinds of soil catenas: (1) those running the length of divides; (2) those running the length of the interfluves that branch off from the divides; (3) those starting at a divide running down stream drainageways; and (4) those running approximately normal to the third. Each of the four sequences may prove to be characteristic of a soil cover pattern. A

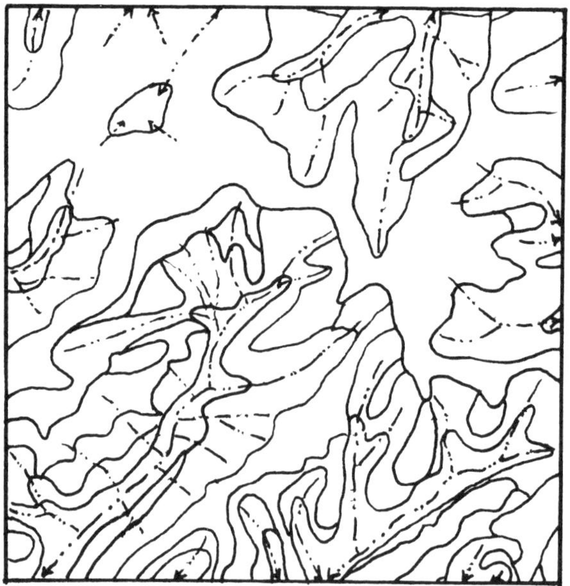

Figure 4.25. Soil cover pattern illustrating openness (external drainage) except for a small (about 3 ha) soil body in the southwest quarted. (Area is 1 sq mi; Sec 11, T 89 N, R 45 W., Woodbury County, Iowa.) (Worster, Harvey, and Hanson, 1972.)

sequence down-valley of the fourth kind constitutes a fifth and compound sequence.

Figure 4.29 presents results of some studies of changes in mean soil moisture regime indices down-valley. The three examples are from semiarid (Idaho), subhumid (Nebraska), and humid (Missouri) regions. The curves in the upper diagram of Figure 4.29 show relatively little vertical displacement. Considerable vertical displacement was found in plots of data (not shown) from small valleys in southwestern Wisconsin: on a divide the SMRI was 40 (for well-drained upland soils) and within 1 km down-valley the SMRI was 80 or more (for poorly- and very poorly-drained alluvial soils). Some small valleys in the Spodosol region of Michigan start in large bodies of peat and continue within narrow ribbons of peat for many kilometers, bordered on interfluves by excessively drained Orthents. Each soil landscape has characteristic signatures of soil body sequences.

Figure 4.30 presents normalized data curves for soil bodies traversed by first and second order streams on 4 geomorphic surfaces in northwestern Iowa (Hole, 1983). The data illustrate that (1) first order streams cross steeper soil bodies than do second order streams, and gradients increase with age of geomorphic surface; and that (2) soil bodies crossed by first order streams are less moist than those crossed by second order streams, and with increasing age of geomorphic surface, the soils of both groups become less moist.

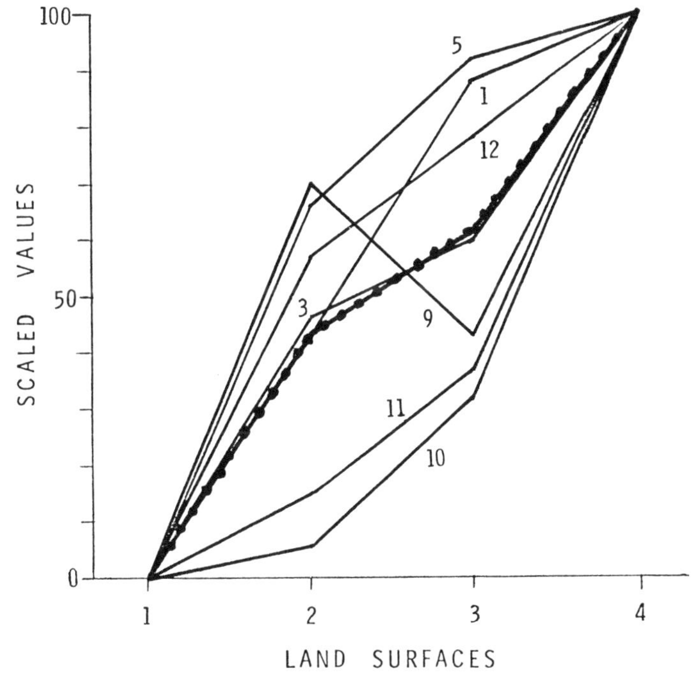

Figure 4.26. Normalized soil cover data curves for a chronosequence of four geomorphic surfaces in northwestern Iowa: (1) Cary moraine (14,000 years old); (2) Tazewell moraine (20,000 years old); (3) Iowa erosian surface (< 30,000 years old); (4) Sangamon paleosol surface (> 70,000 years old). 1 = total length of stream drainageways per quadrat (1 sq mi). 3 = total relief per quadrat. 5 = ratio of mean number of soil boundaries crossed per mile of transect to mean number of contour lines (10 ft contour interval) per mile. 9 = mean degree of soil erosion of soil bodies adjacent to nodes. 10 = same parameter calculated on the basis of all soil bodies per quadrat. 11 = mean percent slop gradient of soil bodies per quadrat. 12 = mean soil moisture regime index per quadrat. (All data based on quadrats.) Dotted curve is the mean of all other curves (Hole, 1983).

Degrees of Complication and Simplification By Human Activity. The complexity of human impact on the soil cover is worthy of careful documentation (Plate 4.2). Cultivation, drainage of wet spots, land leveling, treatment with amendments, and the use of large agricultural machines tend to simplify the soil pattern; microtopography is smoothed, pH and content of plant nutrients and moisture regime are made more uniform. Practice of monoculture and of rotations of a limited number of crops constitutes a dramatic simplification of biotic cover which affects the soil cover. On the other hand, preferential landforming, or fertilization and irrigation of selected areas, creates an artificial heterogeneity of the soil cover. Mixed culture of crops, and the intensive manipulation of soils for the production of special crops

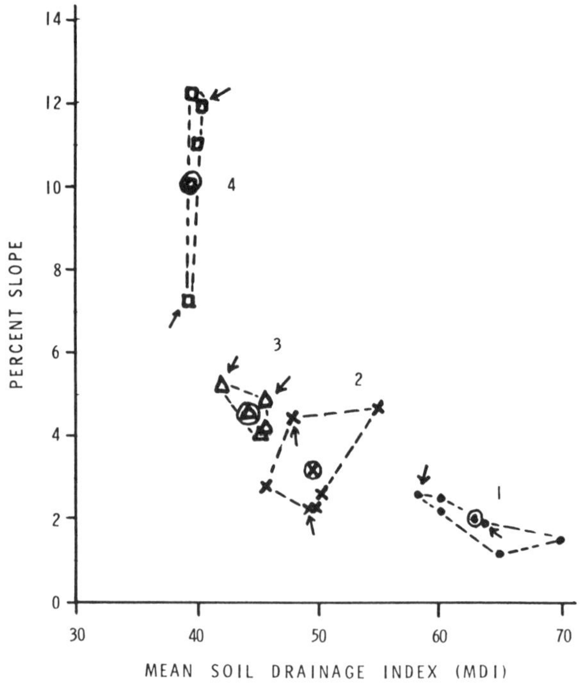

Figure 4.27. A plot of two parameters, numbers 1 and 12, from Figure 4.26 to separate data points (the circled point in each cluster is the mean of the quadrat data points joined by dashed lines) from four distince geomorphic surfaces (Hole, 1983).

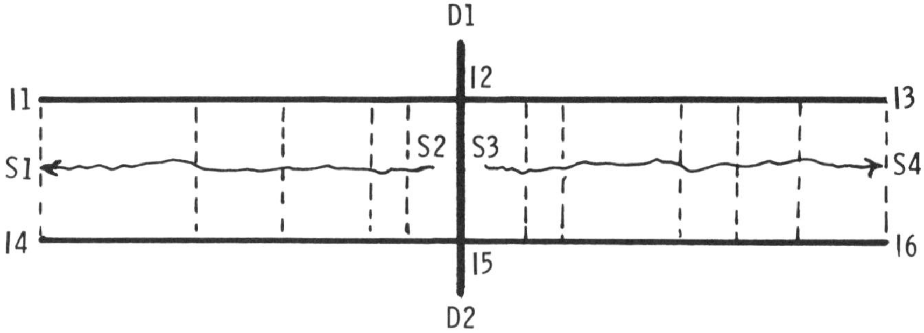

Figure 4.28. Diagram of transects along which different kinds of soil sequences (catenas) may be investigated. 1. D1-D2 represents a divide along which a sequence of soil bodies may be examined. 2. I1-I3 and I4-I6 represent four interfluves that branch off the divide. 3. S1-S2 and S3-S4 are stream drainageways that traverse sequences of soil bodies. 4. The dashed lines are soil catenas from interfluves to drainageways.

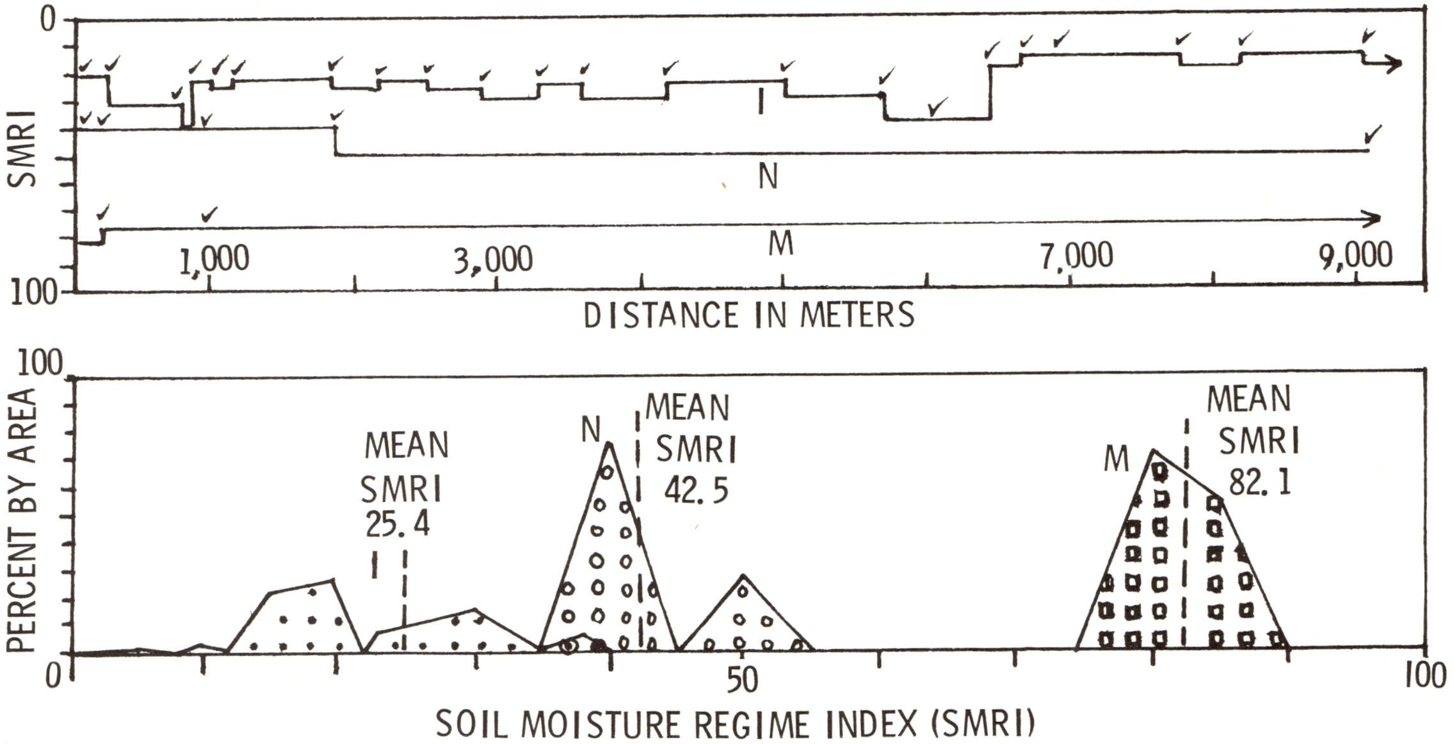

Figure 4.29. Diagram summarizing geographical information concerning distribution of soil moisture regimes in three small watersheds in Idaho (I), Nebraska (N), and Missouri (M). The curves of the upper diagram report changes in soil moisture regime with increasing distance from the divide (far left) downstream (to the right) in the three watersheds. Checks indicate soil boundary crossings at stream drainageways, which are also the places at which catenas of soils on adjacent valley walls to interfluves were examined on soil maps. The lower diagram reports the areal distribution of soil bodies in the same three areas in terms of soil moisture regimes.

such as cranberries, rice, potatoes, and fruit trees creates a complex soil pattern and a new set of soil miroclimatic regimes. Strip-mining and subsequent reclamation of mined lands create other manmade soilscapes, often with their own artificial pedologies. Reclaimed land in coastal regions forms another example of manmade soilscapes. Existing information in the scientific literature on these varied conditions may be brought together by soil landscape analysts to be examined from the perspective of the soil cover pattern.

Differential erosion of soil from elevations and sequential deposition of fractions of the eroded material diversifies soil pattern with respect to landform, materials, moisture regimes, and biotic succession. The percentage

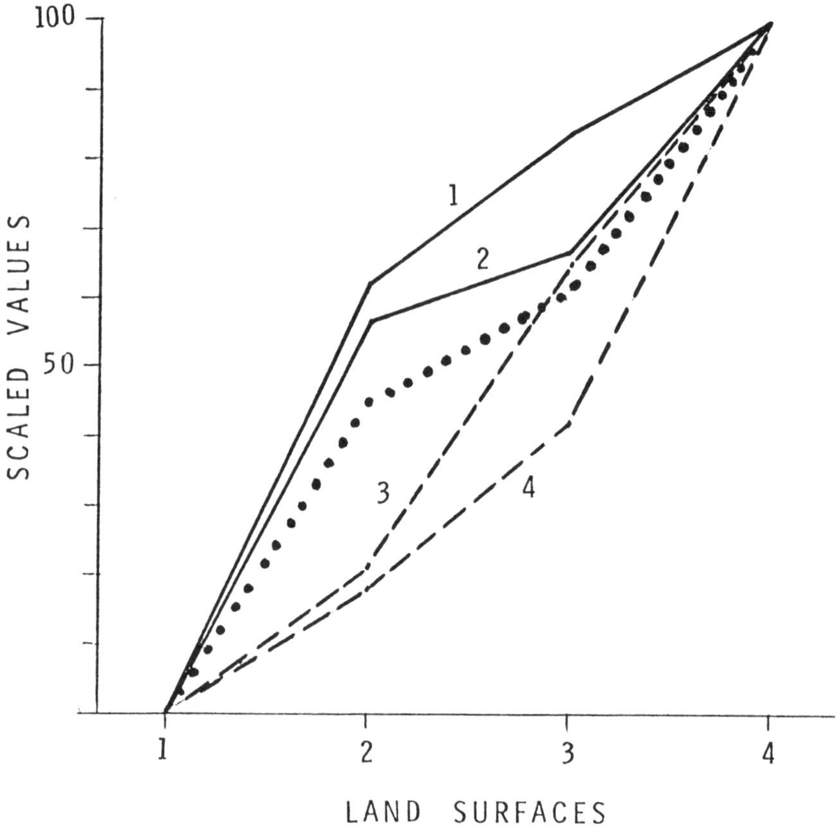

Figure 4.30. Normalized soil cover data curves for a chronosequence of four geomorphic surfaces in northwestern Iowa (see legend, Figure 4.26). 1 = mean soil moisture regime index (SMRI) of soils crossed by first order streams. 2 = mean SRMI of soils crossed by second order streams. 3 = mean percent slop gradient of soil bodies crossed by first order streams. 4 = mean percent slope of soil bodies crossed by second order streams. (Data are based upon quadrats, 1 sq mi each). Dotted curve is the mean of the others.

Plate 4.2 Aerial view of landscape near McAlester, Oklahoma. The irregular patterns of drainage and vegetation contrast with the highly regular patterns formed by mankind's occupation of this landscape.

and equipment is to be found in books and journal articles on soil physics, meteorology, and ecology. Wind gauges, thermometeors, and other meterological instruments are used to monitor movement of air and moisture, such as from over dark plowed fields to adjacent grasslands and woodlands. Microclimatic effects on soil cover by breezes, with and without fog, and from large bodies of water can be recorded. Transport of salt in spray from ocean to shoreland is significant. Movement of water by surface flow across and between soil bodies may be measured by a variety of flumes, weirs, and catchment devices at different scales. Gravimetric and neutron probe methods are available for measuring moisture content of soil horizons, and tensiometric apparati are used to determine rates of movement of water area of local alluvial, colluvial, and cumulic soils in a soilscape may be used as an index of the degree of accelerated soil erosion in the terrain. Once signatures recorded by remotely sensed images are identified as distinctive evidence of accelerated erosion and sedimentation, patterns created by these two processes may be delineated as distinct from those patterns that existed previously. The great detail with which accelerated erosion and deposition features were mapped in the 1930s in the U.S. may well be recaptured as efforts are renewed in the 1980s and thereafter to cope with soil loss and water pollution (Larson et al, 1983).

Kinds and Degrees of Bonding Between Soil Bodies and Other Entities. Ingenious methods are available for measuring interactions between soil bodies through movements of energy and materials. Information about procedures under unsaturated conditions through soil horizons. Efflorescences of salt at footslopes and on crests of knolls are clues to water movement. Analysis of sediment from dust traps and rain gauges provide data on kinds and amounts of dust fall, both wet and dry. Measurement of sediment transport from and storage on soil bodies of valley floors document the dynamics of present-day alluvial soils and of buried pre-settlement soils. Large quantities of plant debris are transported down-valley by floods. Animal ecologists have developed methods of observing local feeding-resting circuits and long-range migration routes of birds, mammals, and insects, many of whose activities are interactive between soil bodies. The soil landscape analyst may find a surprising amount of information on the bonding of soil bodies in the literature under a variety of topical headings.

5

Some Principles of Landscape Mapping

It is difficult to conceive of a landscape outside of the context of maps, or map-like representations. Even if one does not actually use a physical map in the examination of a landscape, it would seem inevitable that the information must be mentally organized in some way that must be essentially similar to the ways that we physically organize information for presentation on maps. Analysis of a landscape requires some form of spatial organization of information at hand; normally maps form the most convenient means of representing spatial information.

Cartographic representations of soil information are based upon underlying models that organize and simplify the complex topical and spatial information that describes the land surface. Each model rests in turn upon assumptions and premises concerning the fundamental character of landscape, especially those that address the geographic expression of landscape qualities.

This chapter will identify a series of assumptions regarding the character of landscape, methods of examining landscape properties, and of representing them on maps. An explicit statement of these assumptions will form the basis for progressive examination, testing, then revision, as required, to form more accurate statements and models.

NATURE OF THE PHYSICAL LANDSCAPE

What we learn about the earth depends largely upon the kinds of questions we pose. The kinds of questions we can ask follow in turn not only from the facts at hand, but also from the conceptual framework that we use to organize and structure these facts. Often the concepts upon which we depend so much are implicit rather than explicit, and therefore are seldom discussed or questioned in the same manner that the validity of specific items of fact may be questioned.

As a result, we can gain from an effort identify and explicitly state some of the assumptions and principles used in the study of landscapes, with specific emphasis on concepts that reflect our knowledge of the spatial organization of the landscape. The list that follows is based upon experience, observation, and scientific literature. Many are well established; others are derived from rather limited observation and and data; still others reflect unverified (perhaps unverifiable) notions of how the landscape is organized. They are intended not as statements to be accepted as fact, but rather as a set of propositions to be examined, tested, then revised as required.

1. *Place-to-Place Variation is an Inherent Quality of the Landscape.* At virtually any scale, from one of millimeters to one of kilometers, almost any soil property exhibits changes over distance. Both direct observation and our knowledge that the soil-forming factors vary in time and in space, both individually and in combination (Ball and Williams, 1968; Beckett and Webster, 1971), have shown that place-to-place variability is universal. Different species of soils have been observed to exhibit different degrees of variation in respect to individual properties (Campbell, 1979).

2. Considered over distances of several kilometers, *Soil Properties Display a Mixed Form of Place-to-Place Variation.* Within soil bodies, we expect to encounter *continuous* place-to-place variation; only at distances of a few millimeters or centimeters do we normally observe voids or other discontinuities. However, relatively abrupt changes in the form of variation occurring over distances of several meters, tens of meters, or sometimes greater distances, are observed at the edges of landscape units. Indeed, we expect to observe a discontinuity in respect to at least one property at the edge of a landscape body, although there may be a wide range in the degree of abruptness at the boundary. This basic model (*continuous variation* within landscape units, *discontinuous variation* at boundaries) is a simplification, but one that portrays a genuine feature of the spatial variation of most landscapes. Thus, the landscape is formed from a series of discrete, interlocking, soil volumes of varied size and shape that each exhibit individual spatial properties (see Fridland, 1976a).

3. *Elements of the Landscape form Spatial Systems.* From classical pedology we know of the intra-profile exchanges of energy, moisture, gasses, and minerals between components of the soilscape, and of the close interconnections between the soil profile and vegetation, topography, and climate (Jenny, 1941; Crocker, 1952). In a similar manner, the landscape forms a spatial system (at least within defined regions, most notably the drainage basin) in which *place-to-place* movements occur (Huggett, 1975). These spatial interactions occur by means of throughflow, soil creep, overland flow, mass movements, wind transport, and other processes of erosion, transport, and deposition (Chapter 3). The evolution of the landscape as a spatial system leads to the ordered spatial occurrence of soils that we observe as catenas, *toposequences,* etc.

4. *Systematic Geographic Variation of Landscape Units.* The widespread use of the catena, the soil association, and analogous concepts indicates widespread acceptance of the existence of an order to the arrangement of soil bodies on the landscape. Although it may be extremely difficult to discern this order at times, few scientists could accept the alternative: that the character and placement of soil bodies are the result of randomly distributed processes.

5. *Landscape Units Can be Identified and Delineated.* Landscape units can be delineated in the field, and despite the numerous limitations of the usual topographic, geologic, and pedologic maps, large-scale maps of national resource surveys depict a wealth of information about the landscape.

6. *Landscape Units Exist as Recognizable Entities.* The existence of definable ("natural") soil bodies has been questioned (Van Wambeke, 1966), but the testimony of soil scientists in many countries, expressed both implicitly and explicitly in published works over many years, confirms the broad acceptance of this concept. We must concede, however, that many soil bodies are indistinct, and that others are highly variable.

7. *Landscape Units Possess Observable Geographic Qualities.*

(a) Among the most obvious geographic qualities are size, shape, topographic position, uniformity, surface topography, slope, and depth. The assignment of such geographic properties to landscape units is not random, but reflects the environmental setting and provenance of each soil. Just as the soil profile records the action of the soil forming factors at a specific point, so also do these geographic qualities reflect the action of these same elements acting upon areal units, and upon interconnected assemblages of areal units.

(b) For a given landscape unit, geographic properties may form distinctive identifying characteristics, comparable in uniqueness to properties of the soil profile (Hole, 1953). For example, specific landscape units may be as distinct from their neighbors in respect to sizes and shapes of delineations as they are in respect to texture or thickness of the B horizon.

(c) The geographic qualities of soil and landscape units influence our ability to acquire information about them, and to make effective use of soil resources (Kellogg, 1949; Riecken, 1963; Lyford, 1974; Larson et al, 1983).

8. *The Geographic Properties of Landscapes Reflect the Action of the Soil Forming Factors.* The assemblage of soil bodies within a given region forms a mosaic of interlocking areal units. The sizes, shapes, and diversity of these soil bodies form distinctive evidence of the environmental character and history of the landscape. Therefore, measurements of sizes, shapes, contrast, diversity, etc. of component soil bodies reflects the origins and distinctive character of specific soilscapes.

SPATIAL ORDER WITHIN THE PHYSICAL LANDSCAPE

If place-to-place variation occurred at random, without elements of organization and order, mapping efforts could proceed only with the greatest difficulty because information and experience gained at one location could have little predictive value at new locations. Under such circumstances each mapping problem would be unique because of the lack of a consistent geographic order that can be transferred from previous experience in analogous settings. In practice, we can often recognize elements of spatial organization within the landscape, and thereby define strategies for simplifying mapping problems.

An enumeration of geographic qualities important in examination of soilscapes includes spatial autocorrelation (associations between measurements acquired at nearby locations), spatial intercorrelation (interrelationships between separate properties observed over distance), and spatial associations (recurring geographic patterns among landscape units).

Multivariate Character of the Soil Landscape. For any volume of soil, or for any unit of the landscape, a scientist can make hundreds, if not thousands, of measurements of soil properties (including chemical, physical, and mineralogical charateristics), and of landscape qualities (such as slope or relief). the properties selected reflect the requirements of a specific purpose, and the constraints imposed by cost, time, effort, and access. Seldom can we examine any single measurement in isolation from others, and seldom can we be justified in regarding place-to-place variation of a given property as independent of others. As a result, a given geographic distribution must be recognized as a factor contributing to another pattern, or as the result of interactions of other distributions. For example, the occurrence of soil manganese is related to variations in patterns formed by the interaction of drainage, relief, slope, vegetation, parent material, and aspect. These patterns are in turn formed from interconnections with other landscape elements. Any given distribution can therefore be visualized as the result of superimposition of several contributing factors, and may itself contribute to the distribution of other properties.

Spatial Autocorrelation. Within nominally homogenous landscape units we expect to encounter continuous place-to-place variation of soil properties. Given observations positioned sufficiently close to one another, we usually encounter the presence of positive spatial autocorrelation. That is, measurements taken at one set of locations are found to be similar to corresponding measurements made at locations near to the first set. (Cliff and Ord, 1973; Davis, 1973). For example, moisture content measured at a given location is more likely to resemble another measurement made at a distance of one meter from the first than it is to another made at a distance of 50 meters. The actual distance over which such positive associations are observed varies with respect to the property studied, and with respect to the uniformity of the landscape unit (Campbell, 1978).

If there are abrupt changes in the place-to-place variation of a property, then positive autocorrelation may not be observed among measurements gathered on opposite sides of the discontinuity. Also, even when place-to-place variation is continuous, we may observe apparent discontinuities if the observations are spaced at distances too large to record the continuity of variation. For example, if moisture content in a specific soil is positively autocorrelated over distances of a few meters, but samples are spaced at distance of 50 meters, then we are unlikely to observe strong positive associations between values at adjacent sample positions.

Spatial Intercorrelations. As we observe and measure landscape properties at varied locations we note that individual properties do not vary independently, but may tend to vary together. As a simple example, Table 5.1 shows a correlation matrix calculated from samples collected in two eastern Kansas Mollisols. The strongest associations here are between gravel and sand content, and between gravel content and pH. These data illustrate, then, the existence of assocations between individual properties, such that the value of one might be used as an indication of the value of another.

These values have been calculated from samples collected from two separate mapping units, without regard to the identities of the two units. If the samples from the two mapping units are considered separately, and used to calculate two separate correlation matrices, the results reveal the contrasting nature of the intercorrelations present *within* the two soil units (Table 5.1). The values for the Pawnee series show a negative relationship between sand content and silt content, and a positive relationship between gravel content and silt content. These values are calculated from 50 samples positioned within the Pawnee; it can be shown that this general pattern of intercorrelation exists within the Pawnee, although the values of individual correlations change as different sets of samples are selected. However, if samples are collected from the adjacent Ladysmith series, associations between these properties change completly (Table 5.1). A new set of interrelationships is encountered.

Table 5.1 Correlation Matrices for Properties of Two Soils

	gravel	sand	silt	pH
		Ladysmith		
gravel	1.00			
sand	0.02	1.00		
silt	-0.05	0.32	1.00	
pH	0.20	-0.14	-0.18	1.00
		Pawnee		
gravel	1.00			
sand	0.50	1.00		
silt	-0.58	-0.34	1.00	
pH	-0.10	-0.08	0.26	1.00
		Combined Pawnee and Ladysmith Samples		
gravel	1.00			
sand	0.64	1.00		
silt	-0.38	-0.14	1.00	
pH	0.33	0.07	0.03	1.00

These examples use only a few easily measured soil properties as a means of illustrating the existence of interrelationships between soil properties, and the changes in interrelationships that occur as different soils are considered. The same kinds of relationships exist between other sets of properties that are of greater significance in mapping, and in management of soils.

Associations of this kind may be the result of indirect physical processes (e.g., texture and color may be consistently associated because of the influence of texture upon drainage, and the consequent influence of drainage upon hue). Or, apparent association may simply be a logical consequence of the measurement system (e.g., because sand, silt, and clay content sum to 100%, any one is predictable from the other two).

Knowledge that co-variation exists as a natural characteristic of landscapes forms the basis of the use of surrogates and proxies. Most mapping endeavors rely upon the use of relatively easily observed properties as substitutes for, or indications of, other characteristics that may not be as easily observed or measured. Once the relationship between the easily observed property is reliably established, it can then be used as an aid in mapping variations in less easily observed characteristics.

In fact, it can be recognized that the concept of a general purpose soils map, and the strategy of dividing the landscape into discrete spatial units, have meaning only if one accepts the presence of spatial co-variation of sets of soil and landscape properties. Many of the most difficult soil mapping problems occur when important management properties cannot be related to morphololological properties unobservable in the field. For example, in north central Virginia, the Elioak and Tatum soils differ in productivity, but are almost identical with respect to characteristics observed in the field, so accurate mapping of behavioral properties is very difficult (Wilson et al, 1983). In a research context, study of co-variation of sets of soil properties may lead to identification of the physical and chemical processes that link the properties, and thereby lead to an improved understanding of soil and landscape evolution.

Spatial Associations. Just as spatial autocorrelation and spatial cross-correlations represent forms of geographic ordering of variation within landscape units, so do spatial associations of adjacent landform units form a kind of systematic geographic structure to the arrangement of discrete units within a region. Different occurrences of a specific soil series, for example, are found in similar topographic positions, develop on the same or similar parent material, and may support similar natural vegetation. Because of the consistent ecological settings of specific landscape units, the landscape as a whole possesses a consistent geographic structure. Individual landscape units tend to occur with characteristic sizes, shapes, slopes, and topographic positions, and to consistently border the same kinds of landscape units as neighbors.

Table 5.2 illustrates the tendency towards order in the arrangement of soil units on the landscape. It shows the number of boundary segments shared by the 18 mapping units represented on a sheet of the Shawnee County, Kansas, soil survey (Abmeyer and Campbell, 1970). For the compilation of

Table 5.2 Boundary Segments Shared by Delineations of the Soil Survey of Shawnee Country Kansas

	An	Bk	Br	Dm	Dw	La	Ld	Me	Mp	Pa	Re	Sk	Sv	Vn	Kb	El	Mm	Em
An	—	1	1	0	0	0	0	8	0	7	0	0	1	0	0	0	0	0
Bk		—	2	0	1	4	7	24	1	31	1	4	2	2	2	2	0	0
Br			—	1	0	0	0	4	0	1	2	0	6	0	5	0	0	0
Dm				—	0	1	1	3	0	0	0	0	0	0	1	0	0	0
Dw					—	0	0	0	0	0	0	0	0	0	0	0	0	0
La						—	5	23	0	14	0	0	15	0	0	4	0	1
Ld							—	16	0	22	0	0	0	0	0	7	0	0
Me								—	0	16	6	0	11	0	0	7	2	1
Mp									—	5	0	0	0	0	0	0	0	0
Pa										—	0	7	9	0	0	1	0	0
Re											—	0	1	0	1	0	0	0
Sk												—	0	0	0	0	0	0
Sv													—	0	0	5	0	1
Vn														—	0	2	0	0
Kb															—	0	0	0
El																—	0	1
Mm																	—	0
Em																		—

Source: Compiled from portions of sheets 31 and 32 in Abmeyer and Campbell (1970).

Key to soils:

- An: Alluvial land
- Bk: Breaks-Alluvial land complex
- Br: Broken Alluvial land
- Dm: Dwight-Martin silty clay loams
- Dw: Dwight silty clay loam
- Ld: Ladysmith silty clay loam
- Me: Martin silty clay loam
- Mp: Morrill-Gravelly land complex
- Pa: Pawnee clay loam
- Re: Reading silty clay loam
- Sk: Shelby clay loam
- Sv: Sogn-Vinland complex
- Vn: Vinland silty clay loam
- Kb: Kennebec silt loam
- El: Elbrook silt loam
- Mm: Morrill clay loam
- Em: Elmont silt loam

this table, each individual boundary segment, regardless of length, is counted as a single unit. Entries in the matrix then show the *number* of contacts between pairs of neighboring delineations. Entries of "0" designate pairs of mapping units that do not neighbor each other on this map sheet; high values indicate those pairs of mapping units that tend to neighbor each other consistently on this map sheet. The results reveal a clear order to the arragement of soil units on the landscape. Specific soil units are shown to prefer certain soil units as neighbors, and to reject others. This effect is, of course, a consequence of the ordered spatial effects of the action of the soil forming factors on the landscape, primarily (in this instance) in response to topography and geology.

Recognition of this order is a significant vehicle for understanding patterns of soil variation, and for effective design of composite mapping units (see Chapter 8). In addition, some patterns may have significance for soil management in agriculture, especially if contrasting soils are consistently found adjacent to one another.

Proxies/Surrogates. The use of proxies in landscape studies exploits a knowledge of the co-variation of selected sets of properties, using the more easily observed properties (the proxies, or surrogates) as substitutes for those more difficult to observe. We cannot observe all properties at all locations, so the spatial interrelationship defined between a proxy and another property permits economical mapping of those more difficult to observe directly. Typically, morphological properties, and topography, slope, and vegetation form the most frequently used proxies for soil studies.

The use of substitutes is not always feasible, as it is not always feasible to define close relationships between the property of interest and other more easily observed properties. Or, the relationship may be valid only for a restricted geographic region, and may have varying accuracy within any given region. And, of course, the effectiveness of the concept depends upon the initial accuracy of the definition of the relationship between a property and a surrogate, and thereafter the astute application of the relationship.

The Ubiquity of the Random Element. Some of the preceding material could be interpreted to construct a model of the landscape characterized by complex, but ordered, variation of landscape elements and properties. Often such order is present, but usually it is accompanied by unordered variation. Random elements can be present in our observations due to errors in measurement, defects in sampling, and coarseness of measurement scales in relation to the scale of variation (Mandlebrot, 1983; Burrough, 1983a, 1983b).

Furthermore, the existence of place-to-place continuity may be observable only over distances so short that its observation by means of the usual sampling procedures is impractical. Some properties may display continuity over distances of meters or several tens of meters, whereas others may vary at such high spatial frequencies that place-to-place continuity would be evident only at sample intervals of a few centimeters. Under such circumstances a relatively coarse spacing of samples might introduce an element of randomness into the observed geographic pattern. In addition, temporal variation may be interpreted as spatial variation. For example, an effort to measure moisture content immediately following a rainfall might introduce temporal error because samples cannot all be collected simultaneously. As a result, changes over distance might be mixed with changes over time, with the result that efforts to interpret the meaning of the observed pattern would encounter error and disorder.

MAP REPRESENTATIONS OF THE PHYSICAL LANDSCAPE

Map representations of the landscape vary greatly in accuracy, legibility, precision, and other qualities. The geoscientist-cartographer almost always finds that compromises and sacrifices must be made in compiling and preparing maps of the landscape. For example, there are obvious trade-offs between legibility and the amount of detail that can be portrayed on a map; finding the appropriate balance is a difficult task, and one that is often completed without explicit recognition of the key issues and concepts.

This section presents an outline of some of the most important map qualities, and their role in the preparation and use of landscape maps. Many of these qualities have been implicitly accepted as the basis for preparation of landscape maps (including pedologic, geologic, and topographic maps) for many years. There is merit in the explicit recognition of their role in the making of maps of the landscape, as such recognition leads to a better understanding of the design and logic of these maps.

Homogeneous Mapping Units. An important tradition in the mapping of landscapes is the use of homogeneous mapping units. This conceptual model represents discontinuous variation between units in an explicit manner (i.e., the line symbols that separate parcels approximate the position and form of discontinuities on the landscape). Variation within units is not symbolized, nor is it explicitly acknowledged in most mapping conventions. Soil surveyors recognize the presence of internal variation within mapping units, as do many map users, yet because of the absence of explicit symbolization or acknowledgement of such variation, most mapping units are shown as homogeneous bodies. The map maker who follows in this tradition is therefore constrained to define mapping units to be as uniform as possible, and to be consistent from one parcel to the next, because the usual mapping models permit few devices to portray variations that may in fact be present.

The clearest statements of this approach to defining mapping units are those of Webster and Beckett (1968) and Butler (1980). They explicitly state the principle that the most useful mapping units, from the perspective of the map user, are those that have the smallest internal variability, as these units permit the map reader to make the most accurate predictive statements concerning actual conditions encountered on the ground within the mapped area.

Composite Mapping Units. Composite mapping units are those mapping units intentionally defined to include dissimilar pedological units. The map maker has been forced to temporarily abandon the goal of defining homogeneous mapping units for practical reasons. Limitations of mapping scale and intensity in relation to complexity of the soil pattern compel the map maker to combine several soils, known to be pedologically dissimilar but to occur together in a geographically intricate pattern, within a single mapping unit. The soil scientist therefore has decided, for specific areas on his map, to abandon the principle of uniform composition of mapping units in favor of one of legibility. If composite mapping units are used, not all mapping units need to be defined as composite units, but only those that apply to the complex occurrence of diverse soils as mentioned above, or possibly those that describe soils of marginal agricultural significance. It should be noted that the principle of explicit description of mapping unit content (see below) is especially important for composite units because of their diversity. As a result, the soil scientist should make an effort to describe to the reader the kinds of soils that occur within composite mapping units, their relative proportions, and if possible their pattern of occurrence (see Fridland, 1976a). There is a long tradition of the use of composite mapping units by pedologists and by others (see, for example, Nygard and Hole, 1949; Christian, 1959; Varnes, 1974; and Robinove, 1981). Chapter 7 describes specific kinds of composite units.

Precision. Precision refers to the level of detail represented on the map. Two forms of precision are relevant. Taxonomic, or informational, precision refers to the specificity of mapping units; highly specific units are precise, and more general mapping units (especially composite mapping units) are less precise, because they convey to the reader less specific information concerning conditions to be encountered on the ground.

High precision is usually a desirable attribute. However, high mapping precision often is associated with higher compilation costs, with large mapping scales, or increased map complexity, so it should not be accepted as a quality to be pursued at any price.

Spatial precision denotes the fineness of the spatial subdivisions; small parcels accompany high spatial precision. It is probably true that high spatial precision generates increased error (at a given level of effort), so there may be a trade-off between accuracy and spatial precision. Selection of an appropriate level of precision is a joint decision of the geoscientist-cartographer and the map user, but once selected, both taxonomic and spatial precision should be consistent within a given map or map series.

Accuracy. Accuracy refers to the degree of correspondence between the map and the actual conditions observed at specific points on the ground. If we visit a set of sites within a mapped region and they all correspond closely to the descriptions for corresponding map units, the map is said to be accurate.

Accuracy can be controlled by two factors. One is the precision of boundary placement. If all boundaries between parcels are placed in correct positions, a map would be completely accurate. However, because of generalization, positional error, and the existence of parcels too small to delineate, all maps inevitably include errors in boundary location. These errors can in part be compensated for by information in mapping unit descriptions; precise specification of the actual composition of mapping units permits the reader to know what to expect, and to anticipate the presence of impurities in areas of complex topography.

A second means of controlling accuracy is to vary the taxonomic detail used to define mapping units. A rather broad definition may eliminate errors that occur when more precise definitions are used. The scientist and the map user must decide together if the increase in accuracy compensates for the loss in specificity. Because increases in precision inevitably generate errors, presumably there is a point where there is a joint optimum.

Consistency. The usefulness and overall accuracy of a map is closely related to the consistency of application of mapping unit definitions to the landscape. Because mapping units cannot correspond exactly to taxonomic units, mapped parcels must include areas of soils foreign to the predominant soil of the mapping unit; often these inclusions are not explicitly identified in the legend. The principle of consistency requires that parcels of any given mapping unit should be composed of a consistent mix of taxa. A map reader who examines parcels of a specific category as depicted on the map expects to encounter a consistent suite of soils on the ground within the boundaries of each parcel. This form of consistency could perhaps be referred to as "taxonomic," or "informational" consistency.

A second form of consistency pertains to the sizes of parcels delineated on the map ("spatial consistency"). A map reader expects that variations in parcel sizes on the map reflect actual variations in the texture and fabric of the landscape as observed on the ground. (This is true even though the reader should not expect an exact one-to-one correspondence between mapped parcels and actual units on the landscape.) Thus, the maker of the map must achieve a consistency in the level of detail depicted on the map, and a consistency in complexity of the map pattern in relation to complexity of the landscape pattern. If, in some instances, it is desirable to depart from this objective, perhaps to use greater detail in mapping soils of higher productivity, the reader should be made aware of the differences in detail used for separate mapping units, or separate portions of the map.

Explicit Description of Mapping Units. The reader of a map is entitled to information known to the map maker concerning the composition of mapping units. The limitations of scale and legibility prevent attempts to exactly represent the variations present within a given landscape. Many users of soil maps have an informal knowledge of the limitations in the accuracies of soil maps, including knowledge of the uncertainties of boundary placement and the existence of unnamed inclusions within mapping units. The principle of explicit description of mapping units specifies that the map maker should convey to the map user, in a formal, explicit manner knowledge of the actual compostion of mapping units. This information can be provided either in the map legend, or in the written descriptions of mapping units.

SOME FUNDAMENTAL DICHOTOMIES IN LANDSCAPE MAPPING

Any map maker must select from a menu of established strategies and traditions in organizing and presenting soil information in map form. Often his choices are determined in part at least by the context in which he is preparing the map, so he may not have a free choice of the various alternatives. He must accept the format in which he is working and use it to its maximum effectiveness. However, the user of a map is in a quite different position: a specific map may have been prepared for a purpose, or in a context, quite different from the one at issue at a given time. The strengths and weaknesses of specific maps cannot be understood without some knowledge of the main approaches that have been developed for organization of cartographic data. The following paragraphs present capsule summaries of some of the main dichotomies in the construction of landscape maps.

The Map as Data Storage vs. The Map as A Communication Device. Maps differ greatly in the intricacy of the patterns (level of precision) they show. Some variations in detail are due to differences in the complexities of the mapped areas; other variations are due to differences in the degree to which a map faithfully depicts detail. Those that sacrifice visual clarity for fine map detail can be said to approach the ideal of a map as information storage; the map design is tailored for maximum information content, with little regard for ease of assimilation of the information.

Other maps, however, may make concessions to the reader-selected forms of complex detail may be omitted in an effort to render the map pattern in a manner that permits convenient interpretation of the fundamental pattern. The omitted detail that is sacrificed for the sake of legibility may be small in relation to the benefits gained by improved comprehension by the reader. The simplified map may be significantly more effective, especially if the intended audience has an interest in broad-scale soil patterns, rather than conditions at specific sites.

These two approaches to map making are in fundamental conflict with one another, since attainment of maximum information content (the ideal of the map considered as "data storage") requires maximum complexity, and maintaining high legibility (the ideal of the map considered as a "communication device") requires loss of information. Usually a given map contains elements of both traditions, although it may favor one approach over the other.

Often the strongest criticisms of maps arise from conflicts between these two strategies. Incorrect placement of boundaries, or the presence of inclusions within mapping units, are often perceived as serious errors by those who implicitly have accepted the notion of the soil map as information storage (i.e., as a vehicle for conveying precise, accurate, information for each site depicted on the map). Such maps are of course prepared with regard for graphic and logical clarity and therefore include numerous concessions to economy and legibility. They therefore must sacrifice some detail; such are the characteristics of maps designed for effective communication to the reader.

Uniform Mapping Units vs. Composite Mapping Units. In the ideal, mapping units should be defined to be as uniform as possible in respect to the properties of interest. In reality, complex landscape patterns often prevent legible representation of small delineations at the usual mapping scales, and the limitations of time and cost prevent mapping of areas at the same fine level of detail that might be desired. As a result, it is often necessary to define mapping units composed of two or more different soils.

Ideally, such mapping units might be formed by combining units that are selected to be as pedologically similar to one another as possible. This strategy preserves, insofar as may be possible, the notion of homogenous mapping units. Webster and Beckett (1968) present an argument for the definition of uniform mapping units. On the other hand, the soil pattern may dictate a different strategy for defining composite mapping units. The alternative is to combine those soils that tend to occur together on the landscape in consistent patterns, even though they may be quite unlike each other pedologically. Robinove (1981) outlines a mapping strategy based upon the premise that that uniformity is not generally present in nature to the extent that it can be effective as a basis for landscape mapping, and that mapping units should be defined to recognize this fact by describing complexity that is present, rather than searching for an (in his view) illusory uniformity. Robinove advocates definition of composite mapping units, with complete description of their content, as a means of representing natural landscape patterns. (His ideas are based upon application to remotely sensed data of some of the ideas expressed in other contexts by Christian (1959) and others, but his paper is in effect a statement of a strategy applicable to landscape mapping in general.)

Special Purpose vs. General Purpose Maps. One of the most basic decisions is the selection of scope of content and of intended audience. The general purpose map addresses a broad audience; it employs a classification that attempts placement of soils in a comprehensive taxonomy addressing a broad range of soil characteristics and uses. In theory, taxa in such a classification can be interpreted with respect to any specific use that may be at issue, even though such might not have been explicitly considered at the time of compilation. Advocates of general purpose maps maintain that such classifications address fundamental properties of soils–properties that can be correlated with other soils in analogous settings. Because of the difficulty of designing an effective general purpose classification for large areas, they tend to be unwieldy in a mapping context.

Advocates of special purpose mapping strategies state that in practice, no general purpose classification is likely to be statisfactory for all potential uses,and that resources may be more effectively employed in making maps and classifications tailored to specific geographic regions and specific purposes. Disadvantages of this approach, are of course, that a set of incompatible special-purpose classifications prevents convenient transfer of experience gained in one area to analogous regions, and that the same area must be continually re-mapped as additional management options are considered, or as new crops are introduced.

6

Geographic Concepts in Soil and Landscape Studies

SOIL GEOGRAPHY: TWO TRADITIONS

Boulaine states that "pedology is one of the domains of knowledge for which the geographic approach is absolutely necessary" (1974, p. 7, translated). There is, of course, a large body of knowledge existing under the label, "soil geography"–a knowledge that has both benefited from and contributed to the broader inquiries of pedology itself. This chapter is devoted to an account of another form of geographic knowledge of soils, a line of inquiry that dates to the early days of soil science, yet one not normally included in the scope of the usual definition of soil geography. This second soil geography is in fact truly "geographic" in content and perspective even though it is usually not so named; we could refer to it as "the *other* soil geography." There are then *two* soil geographies–one in the familiar meaning of the term, and the other, a more esoteric and specialized, but equally significant body of knowledge.

The more familiar form of soil geography focuses upon the study of geographic distributions of specific soil taxa, with emphasis upon their character and genesis, their interrelationships with the environment and mankind, and their global and regional locations (Table 6.1). This is the soil geography of dictionary definitions, textbooks, and university curricula. Soil distributions are examined at rather broad scales, often at global or continental levels. The emphasis is upon the soil profile as a fundamental unit of examination. The "geography" of this soil geography focuses on specifying the relationship between location on the earth's surface and observed soil characteristics, and of explaining the logic of these relationships. A concise statement of some of the key elements of this soil geography can be found in the opening paragraph of V.V. Dokuchaev's *Russian Chernozem*:

> Precise knowledge of the geography of Chernozem is prerequisite for establishing its origin; its geography is related to the distribution of certain wild plants and animals in Russia; . . .the geography of Chernozem is genetically related to the climate of the country and partly also to its recent geological history. (Dokuchaev, 1883, p. 1).

Other statements of this perspective are found in Putnam (1951), Duchaufour (1970), Bridges (1978), Bunting (1965), Cruickshank (1972), and the literature review by Bridges (1977).

Table 6.1 Content of Traditional Soil Geography

Global placement of major soil taxa
Soil genesis
Interactions between soil and environment
Interactions between soil and mankind
Soil survey and mapping
Paleosols
Soils—Geomorphic relationships

(This list is illustrative rather than exhaustive.)

Table 6.2 Content of the Other Soil Geography

Definition of Fundamental Areal Units of Soil
Areal Properties of Individual Soil Bodies
Geographic Properties of Individual Soil Species
Genetic Links Between Neighboring Soils
Cartographic Representation of Soil Patterns
Generalizations of Soil Patterns
Examination of Soil Boundaries
Origins of Soil Patterns
Place-to-Place Variation Within Landscapes and Soil Units

(This list is illustrative rather than exhaustive.)

The second soil geography, although almost as old as the first, has persisted as a field of inquiry through efforts of scientists in several nations. Although it overlaps with the traditional content of soil geography, the findings and concepts of this field of inquiry are seldom explicitly acknowledged in the texts and scientific literature that form the mainstream of modern soil science.

The "geography" of this soil geography denotes those properties that pertain to soil bodies specifically as *areal* entities possessing characteristic sizes, shapes, volumes, slopes, and internal variability (Table 6.2). The scale of study is usually local or regional. Although it does not ignore the soil profile, the emphasis is directed towards study of the occurrence of soil units on the landscape in characteristic areal forms and patterns, and the significance of these patterns for mankind's use of soil. There is a long history of interest among soil surveyors in this aspect of the soil landscape, although usually it assumes incidental, rather than primary, significance in conventional soil science and soil survey (Table 6.3). The essence of this approach, and its contrasts with more conventional studies, can be illustrated by Jones' (1959) dissatisfaction with the traditional emphasis upon the soil profile as the basic element for conveying knowledge of soil:

> It is fundamentally unsound to accept the soil profile as an expression of soil cover, since soil is essentially a part of the landscape and a soil classification must contain within itself an inherent reflection of three dimensions (p. 200).

The practical significance of this perspective is evident in Riecken's (1963) work concerning "horizontal applications of technology" to agricultural uses of soil resources, and in Lyford's (1974) assessments of the sizes and shapes of soil bodies in soil survey and resource management in forested areas of New England. Larson et al describe situations in Minnesota where the complex spatial arrangements of soil units inhibit effective use as well as conservation practices: "Because of the complex nature of the topography it is difficult to use a variety of management techniques, and indeed usually this does not occur. Rather, the farmers arrange their fields in large rectangular blocks and manage all soils alike" (1983, p. 460).

In the United States, soil surveyors have commented on the current lack of information concerning the geography of map units (Wilding and Miller, 1979, p. 88) and have recommended that future policies emphasize the development of ". . . more *not* less comprehensive descriptions of mapping units including kinds, amounts, and spatial distribution of component soils" (Wilding and Miller, 1979, p. 92). These kinds of statements from practicing soil surveyors emphasize the concrete, practical dimensions to a subject that may appear initially to be rather abstract and theoretical.

The distinction between these two soil geographies has been described by Fridland (1976a, p. 1) as the difference between focus upon taxonomic units and focus upon territorial units. He notes that this duality "is a feature of all the geographical sciences," and cites botany, zoology, geology, and pedology as examples. It seems clear that half of this duality has been neglected in much of the formal research agenda of pedology and geography, so the purpose of this chapter is to outline in rather broad scope the character of the second soil geography, examine its history, content, basic concepts, and relationships with other forms of pedological knowledge.

CONCEPTS OF SOIL VARIATION

The kind of knowledge that one can acquire about soil depends not only upon the facts at hand, but also upon the research tradition that one accepts as an organizing framework for ordering these facts. For example, the concept of soil as a natural body characterized by distinctive sequences of genetic horizons produced by environmental forces is only one of several alternative views of soil (Simonson, 1968). If one ignores this organizing framework and accepts an image of soil as simply a reservoir of nutrients, or

Table 6.3 **Geographic Properties of Soil Bodies Recorded by Soil Surveyors**

Property	Observation	Source
Size	"Isabella sandy loam occurs in comparatively small scattered areas throughout the country."	Wildermuth and Kraft (1926), p. 15
	"Areas of the Chelmer series are too small to be fully exploited"	Thomasson (1971), p. 64
Shape	"Bellfontaine sandy loam occurs in small irregular areas scattered throughout the country, largely as narrow, broken strips along stream ways"	Wildermuth and Kraft (1926), p. 19
	"because these soils occur in irregular-sized, irregular-shaped, closely associated areas, agricultural practices are similar on various members of the group."	Rogers et al. (1946), p. 20
Homogeneity	"They also show a lack of uniformity, having textural and other variations within very short horizontal distances"	Veatch et al. (1927a), p. 10
	"It occurs in fairly large uniform areas in the western part of the country and in smaller bodies on smooth rounded ridges which rise conspicuously above the level lake-bed plains of the eastern part."	Veatch et al. (1927b), p. 19
	"The Cottam series forms a reasonably uniform mapping unit"	Thomasson (1971), p. 64
Topographic surface of the soil body (relief)	"The surface ranges from undulating and smoothly rolling to rough and hilly. In places it is choppy, with high knolls, sharp ridges, steep slopes, and rounded depressions."	Wildermuth and Kraft (1926), p. 19
	"The surface ranges for fairly level, moderately choppy, and ridgy to rolling."	Ibid., p. 15
Topographic position	"These soils occupy depressions throughout the glacial till, generally small and irregular-shaped and closely associated with areas of well drained and imperfectly drained soils of the uplands. This makes special handling and cropping impractical, except in a few areas."	Rogers et al. (1946), p. 28
Geographic pattern of a specific series (arrangement)	"The areas of Carrington silt loam are rather widely scattered over the eastern half of the country, but the total acreage is small."	Gray and Henderson (1928), p. 22
Slope	Commonly documented by national soil surveys as an important characteristic of soil units.	
Boundaries	"The boundary of the Torbryan series to deeper soils of other mapping units is usually marked by a break of slope."	Clayden (1971), p. 123

as a collection of weathered minerals, it is difficult if not impossible to accurately predict their properties and geographic occurrence.

In a similar manner, our conceptual models of soil variation influence our studies of soil geography. If we accept an incomplete model of soil variation, we tend to pose inappropriate questions, and thereby receive unsatisfactory results. It is appropriate therefore to examine alternative models of soil variation as a means of acquiring a perspective for the study of the geography of soils.

Here the term "geographic" identifies concepts or characteristics that pertain inherently to areas (or volumes) of soil, and their arrangement on the landscape. In contrast, "non-geographic" properties define soil characteristics that do not necessarily require consideration of areal units, that can instead be observed at a single point on the earth's surface. Thus, specification of the size or shape of a soil unit, or of its place-to-place variability, reflects an interest in the areal ("geographic") qualities of soil, whereas an exclusive focus upon the soil profile, observed at a single point, does not. This distinction forms the essence of the study of soil landscapes.

As an example of a concept of soil that lacks the geographic dimension, consider an engineer's definition of soil: "Soil is a natural aggregate of mineral grains that can be separated by such gentle mechanical means as agitation in water" (Terzaghi and Peck, 1948, p.4). Because such a definition includes no mention of horizonation, or of genesis, it provides no basis for predicting variations with depth (of a specified property) or for forecasting place-to-place changes. A person who relied exclusively upon this kind of definition would be unable to explain variations in samples collected at different depths or locations. It would be necessary to turn to another definition of soil—that provided by pedology—to acquire the perspective required to explain the observed changes. In practice, of course, engineers do at times turn to pedology and geology for this kind of information, even if they tend to rely mainly upon parent material as a guide to predicting engineering properties.

Primitive Models of Soil Variation. The most primitive concept of soil variation is to perceive the soil landscape as pedologically homogeneous, without meaningful variation. This notion may perhaps represent the thoughts of young children, or of individuals isolated from direct experience with the environment, but probably it is not an important or widespread concept. However, hints of such rudimentary concepts of soil variation can be detected in early artistic renditions of landscapes scenes. Western artists who first distinguished the contrast between soil and rock did so by painting soil as a fleece-like covering over geologic strata; a clue perhaps to the primitive notions of soil held by some before explicit scientific inquiry was focused upon soil.

Such rudimentary notions of soil variation should not be confused with the knowledge of soils demonstrated by aboriginal or peasant farmers who, although technologically primitive, may possess sophisticated knowledge of soils and soil ecology. For example, Spencer (1966) describes cropping strategies used during the practice of shifting cultivation in southeast Asia, and the rational for site selection. Although he reports wide variation in the astuteness of their observations and practices, it is clear that the most sophisticated of these groups possess a substantial knowledge of soils and soil variability. Experienced tribesmen can evaluate sites on the basis of soil quality, surface uniformity, exposure, drainage, and slope. The presence or absence of certain native plants is recognized as a diagnosis of suitability for growth of certain crops. Such [indirectly derived] fertility ratings employed by some shifting agricultural groups are said to correlate with physical and chemical analyses of soil properties (see for example, Spencer, 1966, pp. 35–36). In a similar context, Moran (1981; pp. 105–110) discusses site selection by pioneers in the Amazonian rainforests, and cites examples of ecologically perceptive decisions made by colonists who are ostensibly unsophisticated in scientific matters. (It should be noted that both Spencer and Moran also cite examples of unwise or erroneous decisions by both groups.)

Discontinuous Variation. The conventional soil map implicitly employs a model in which soil bodies form discrete, internally uniform, units, with abrupt discontinuities at their edges. The soilscape is depicted as a mosaic of interlocking units with sharp boundaries. Although most scientists recognize that this model is unrealistic in some respects, most national soil surveys have accepted it for routine use, if only as a convenient device for representing a complex reality.

Although many readers of soil maps recognize clearly, in the abstract, that this cartographic model often does not correspond to landscape reality, it is important to understand the nature of this approach because it is so often used as the primary means of portraying our knowledge of soil landscapes. Our vision of the landscape may be formed as much from such cartographic representations as it is from our direct observations in the field; in fact, one might argue that such cartographic representations, through their subtle influences upon our visualization of landscapes, greatly influence the manner in which we acquire and assimilate information in the field. Lyford (1974) has stated:

> "Regrettably there is a tendency among users of maps, soil or otherwise, to consider any delineated area on a map as uniform, especially if it has a single color or a single symbol. All too frequently the accompanying description is not read carefully" (Lyford, 1974, p. 207).

In addition, it must be noted that many of the most enthusiastic critics of soil maps, and of soil mapping programs, are often those who interpret (despite conceptual and practical evidence to the contrary) this cartographic model as a literal representation of the landscape, and expect to encounter in the field the discrete, homogenous units depicted on the map. Therefore, even though the limitations of the discontinuous model of soil variation are well known, it is useful to enumerate some of its deficiencies.

This model is unrealistic because it does not represent internal variability within mapping units; often only a single profile description is presented for each mapping unit, and seldom is there a systematic effort to describe the range in variation of observed properties. Often it is necessary to include, without notice, areas of foreign soils within mapping units. Although the existence of internal variability, and the presence of unnamed inclusions within mapping units has been repeatedly reported in scientific journals, the usual maps and reports seldom state the degree of variation present, [or even acknowledge its presence.]

Seldom are soil boundaries as distinct as the lines used to symbolize them. Although it is possible to identify in nature extremely abrupt boundaries, they are not as common as literal interpretation of a map would suggest. Actual boundaries, of course, exhibit a wide variation in boundary form—variations that are usually not symbolized on soil maps. Thus, this model of soil variation must be regarded as a useful expedient, but one that generally provides only a rough approximation of the actual landscape.

Continuous Variation. A third model of place-to-place variation is based upon the premise that soil properties vary continuously over the earth's surface. A rational for this concept can be based upon the notion of a soil universe (Knox, 1965; Schelling, 1970) consisting of all soils that exist at the present, that have existed in the past, and that will exist in the future. If it were possible to collect the infinite number of soil profiles observations that comprise this universe, then place them in a sequence so that those profiles most nearly similar are adjacent, they would then form a *soil continuum*, and any two contrasting soils (on the continuum) would be connected by a series of transitional soils. The extreme variety and complexity of the soil universe would mean that this continuum would extend in many directions simultaneously, perhaps in a manner analogous to a multidimensional Munsell color solid. In forming soil classifications we subdivide this continuum, usually into arbitrary segments thought to be useful for the study and management of soils. Because we can observe portions of this continuum in nature, we might assume continuity as well for the place-to-place variation of soil properties.

Comments of some soil scientist might support this concept: "individual bodies of soil are seldom set apart from their neighbors by sharp boundaries" (Simonson, 1959, p.152). Van Wambeke (1966) maintains that there are no easily identified, "natural" units of the soil cover, and that those units that we arbitrarily delineate do not correspond to taxa in our classification systems. Jones (1959), Arnold (1964), Walker et al (1968), and Davies and Gamm (1970) have all conducted research based upon the premise that the soil landscape exhibits continuous variation. Jones (1959, p. 198) maintains that "no soil possesses a discrete existence but merges imperceptibly into the soil in juxtapostion with it."

It must be recognized that the presence or absence of place-to-place continuity depends in part upon the scale of one's observations. At a very fine level of detail (e.g., a few centimeters) almost all soilscapes would, for most practical purposes, exhibit spatial continuity; one would not observe notable voids or discontinuites in the commonly measured soil properties. At coarser scales (perhaps with observations spaced at several tens or hundreds of meters) most soil distributions will appear to be discontinuous; one would observe abrupt changes from one measurement to the next, with no apparent continuity.

Therefore, the continuity of spatial variation of soil properties cannot be treated as an absolute, and we cannot claim that either continuous or discontinuous variation is the "correct" model for variation within the soil landscape. Nonetheless, it seems clear that, at scales familiar to most soil surveyors, discontinuities are present in the soil landscape to the extent that they cannot be considered rare or unusual (Webster, 1978). Furthermore, within a single geographic area one property may exhibit (at a given sampling interval) continuous place-to-place variation, whereas another may display continuous variation (Campbell, 1977).

Mixed Variation. Other scientists have applied models of soil variation that recognize the existence of discrete units, defined at their edges by *discontinuities,* but exhibiting *continuous variation within these boundaries.* The abstract argument for this model is based upon the fact that in nature we cannot observe the soil continuum (except in small segments) because sections have been removed (by natural spatial and temporal variation) from their ordered positions in the continuum and placed on the landscape to form the mosaic of soil bodies that we observe as the soil cover. Thus, a collection of pedons from adjacent, or nearly adjacent, positions on the soil continuum form "natural" soil bodies in the sense of Knox (1965) or Schelling (1970). Within its boundaries such a body can be considered to exhibit continuous place-to-place variation; at its edges we observe the sometimes gentle, sometimes sharp, transitions to other soil units that occupy adjacent positions on the landscape, but quite different positions on the soil continuum. Much of the research that applied regionalized variable theory to the analysis of soil variation used a mixed model for variation of soil qualities (e.g., McBratney et al, 1982); the nugget effect observed in the semivariogram accounts for the discontinuities present within the observed place-to-place variation, while the remainder of the curve represents the overall character of the continuous spatial variability. (See also Burgess and Webster, 1980a and 1980b).

This notion is consistent with the observations of Fridland (1976a, pp. 15–16) who writes that "... the soil cover may be regarded as a discrete-continuous formation that is physically continuous, but geographically discrete ...", and "... classificationally it [the soil cover] is liable to be either continuous (with gradual transitions between soils, though closely related soil forms), or discrete (with sharp transitions between soils and very dissimilar neighboring soils)."

If one must accept a single model for soil variation, this one of mixed variation seems to be the one most nearly consistent with our experience, at least at the level of detail usually applied in modern soil survey. However, such a model is realisitic only if we recognize the great variation in the degree of abruptness at boundaries to the point that in some instances a boundary may be difficult to observe, whereas others are easily observed. We should also expect some soil properties to change very sharply across boundaries, while others display very gradual changes across the same boundary.

Ordered vs. Disordered Variation. Examination of the soil landscape reveals another form of mixed place-to-place variation; a mixture of order and disorder are sometimes recognizable at several scales. If we examine a spectrum of soil properties, we can often discern elements of order in the sense that some properties will covary as one moves from one place to another. For example, Table 5.1 presents a variance-covariance matrix for properties measured within the Pawnee clay loam (a Kansas Mollisol). This matrix remains essentially the same for other sets of measurements gathered within the mapped limits of the Pawnee series, although random variation contributes to small differences. However, this set of relationships breaks down as one traverses the boundary into another soil unit, where another set of relationships is established as a more or less consistent pattern among variables within the second soil unit shown in Table 5.1. Thus, we can discern a coherent structure to the covariation of soil properties within one landscape unit, one that is subject to small random variation within the landscape unit, and to major shifts in structure between landscape units.

Another form of order/disorder is present within the landscape. The placement of soil bodies on the landscape is *disordered* in the sense that adjacent segments of the soil continuum may be separated on the landscape by distances of several kilometers, or perhaps even hundreds of kilometers. Yet arrangement of soil bodies on the landscape is *ordered* in the sense that we often find consistency in the geographic placement of soil bodies. Soils A, B, and C may be consistently associated with one another within a region, or soil A may exhibit a characteristic geographic manifestation in respect to size, shape, geographic position, etc. Likewise, specific segments of the soil landscape may display distinctive and consistent spatial characteristics. For example, it may be true that floodplains within a region may tend to be populated by soil bodies at specific densities, exhibit characteristic ranges in pedologic properties, and display distinctive shapes.

APPLICATIONS

A set of examples, each presented here briefly in capsule form, illustrate practical problems that require application of the approach outlined in previous sections of this chapter. None of these problems has as yet been "solved," and in fact some may not have "solutions" in the usual sense of the word, but in each instance the perspective and knowledge of soil geography forms an essential part of understanding the problem and establishing the proper context for its resolution.

Application of Agricultural Fertilizers. Manufactured fertilizers form an important component in modern agricultural systems, and one that has received increasing attention as costs of manufacture and application have increased in recent years. Considerable research has been devoted to understanding the best ways to select appropriate fertilizers and rates of application. Some of the most interesting elements of this problem are the difficulty of uniform application, and the assessment of appropriate rates of application in the context of inhomogeneities present in the soil substratum. Riecken has noted that "crop production is necessarily geographic," and that "realization of the 'technologically possible' from individual soils depends among other things on their spatial features" (1963, p. 53). Yet many of the analyses of this subject have neglected the spatial components that are so important.

An economic analysis by Jensen and Pesek (1962) examined the implications of non-uniform application of fertilizers due to imperfections in manufacture, storage and handling, and application by farmers. Their models examine the effects of varied patterns of application, and the effects of physical segregation of blended fertilizer materials. In one such analysis they report losses of some 5 to 6 bushels per acre for corn raised on the Ida silt loam in Iowa.

Although they conduct a detailed analysis of the effects of spatial variations in fertilizer applications within fields, they assume that the soil itself is spatially homogenous, and thereby neglect to consider another source of place-to-place variation that could magnify some of the effects they observed. The significance of within-field pedological inhomogeneities upon observed crop yields has been clearly documented (Harris, 1915).

Spatial variability within agricultural fields would seem to have important influences upon uses of fertilizers. One is that fertilizer recommendations must be based upon information acquired from soil samples (Peck and Melsted, 1967). The accuracy of this information as a guide to conditions within the field as a unit depends upon the number and placement of samples in relation to the amount and pattern of variation present. (One assumes that the field has been placed within a single soil landscape unit; many fields, of course, are placed across soil boundaries, which will greatly increase the variability.) Even under the best conditions, we can never know beforehand the optimum number and placement of samples for a specific field. If samples should be spaced too closely, they cannot provide independent estimates of soil qualities, and variability can be systematically underestimated (Hills and Reynolds, 1969; Basu and Odell, 1974; Agterberg, 1965; Campbell, 1981). As a result, some prior estimate of within-field variability is essential for reliable estimation of soil properties for fertilizer applications.

Soil Data for Hydrologic Models. Civil engineers and hydrologists form computer similations of the manner in which watersheds respond to a series of rainstorms as a means of learning about hydrologic processes in general, and about specific watersheds. In instrumented watersheds, rainfall measurements have been recorded at intervals by means of raingauges, and streamflow by means of stream gauges placed at intervals along channels within the watershed. Hydrologic models attempt to use the rainfall data, together with a knowledge of the drainage network, topography, soil, and land cover, to forecast the magnitude of the peak flow, the timing of the peak flow at points along the channel, and the total discharge with specified time intervals.

Such models depend upon the accurate segmentation of the drainage basin into hydrologic response units (HRUs), defined to delineate areas of uniform hydrologic behavior. For each HRU, a model estimates the amount of precipitation received, moisture retained by soil and vegetation, moisture that enters the water table, and moisture passed to lower HRUs as surface runoff. Modeling of interactions with adjacent HRUs upstream and downstream leads to estimates of amounts of moisture entering the stream system at specific times and places. By this means the overall behavior of the drainage system is approximated.

A key feature of such models is the accurate, economical, description of HRUs. An ideal HRU would have uniform slope, land cover, and soil substratum, and would therefore behave in a hydrologically uniform manner. Although (in theory) accuracy increases as a drainage basin is subdivided into many, small, unform HRUs, practical constraints require that fewer, larger, and less uniform HRUs be defined. Because of the important role of soil characteristics in determining the hydrologic properties of HRUs, this problem requires examination from the perspective of the soil landscape. It is necessary to examine each drainage basin to define composite units that possess a degree of pedological uniformity, and to consider relationships with neighboring soil units, the effects of slope, etc.

Soil Data for Geographic Information Systems. Geographic information systems record data, coded by geographic location, for geographic units as large as entire states. Typically such systems are designed to display, in a standard format, data pertaining to both natural and cultural landscapes, including information pertaining to soils, drainage, land use, vegetation, population, economic and demographic characteristics, etc. Soil information has been an important part of several such systems.

Among the numerous practical and conceptual problems are those inherent in coding soil data for entry into a data base. Many systems rely upon a network of uniform cells as the fundamental geographical and logical units for the data base. Each cell can record only one soil identity (although some systems may record several characteristics for the single soil identified). From a geographic perspective, the most serious problem is caused by the fact that the usual cell sizes are large enough to encompass portions of several soil delineations even in landscapes of rather low or modest complexity. A cell may encompass (for example) 6 distinct soils, even though only one can be recorded (Wehde, 1982). As a result, an arbitrary rule must determine which soil is to be selected to code the entire cell. Often the areally predominant soil is selected; an alternative is to select the soil that falls beneath a randomly selected dot positioned within each cell. The latter strategy improves the overall statistical accuracy of the data base, but may code individual cells with soils that have very small areal proportions of the entire cell.

Figure 6.1 illustrates the problem. The predominant soil strategy systematically excludes those soils that, by virtue of size or shape, can never dominate an entire cell. The smallest delineations, and those that occur in long, thin, delineations are excluded in favor of those that occur in larger, more compact, units. As emphasized elsewhere in this volume, the sizes and shapes of soil units are by no means accidential properties, but may be intimately associated with pedological properties of great significance to the user of a data base. For example, the long, thin delineations mentioned above could well be those soils associated with floodplains. For agricultural uses, or for use by county planners (for example), the systematic exclusion of floodplain soils would probably render a database useless for many important purposes!

Thus, the distinctive sizes and shapes of soil units control the manner in which a given cell size records the character of the soil landscape. Alternative strategies, such as increasing cell size, or recording soil boundaries in polygon form, may not be feasible for large data bases. One practical solution is to generalize the soil pattern into a simpler form that is designed to preserve the essential features of the soil pattern, by using fewer, but more broadly defined, soil units. Definition of such units requires an intimate knowledge of the geographic qualities of the soil landscape (Chapter 8).

Mapping Soils Formed on the Rome Formation in Southwestern Virginia. As outlined above, the usefulness of a soil map can be considered directly related to the user's ability to make precise and accurate statements about the areas represented by the mapping units (Webster and Beckett, 1968). That is, the usefulness of a soil map depends in part upon the soil surveyor's ability to define uniform mapping units, or to specify the amount and pattern of variation within composite mapping units. The uniformity of mapping units is determined in part by the map legend, the surveyor's ability to delineate units in the field, and the inherent variability present in the landscape. In complex landscapes the degree and pattern of variability assume a significance that can approach the limits of the usual models of soil variability to represent existing variation.

The soils developed from the Rome formation in southwestern Virginia are an example of the practical problems encountered in complex landscapes. The Rome formation (designated the Waynesboro formation in northern Virginia) is of significant areal extent in the limestone valleys of Virginia and nearby states. It has been estimated that the Rome-Waynesboro formation covers over 170,000 ha (419,900 acres) in the eastern Ridge and Valley physiographic province in Virginia (Campbell and Edmonds, 1984). Large areas of the same formation are found in analogous

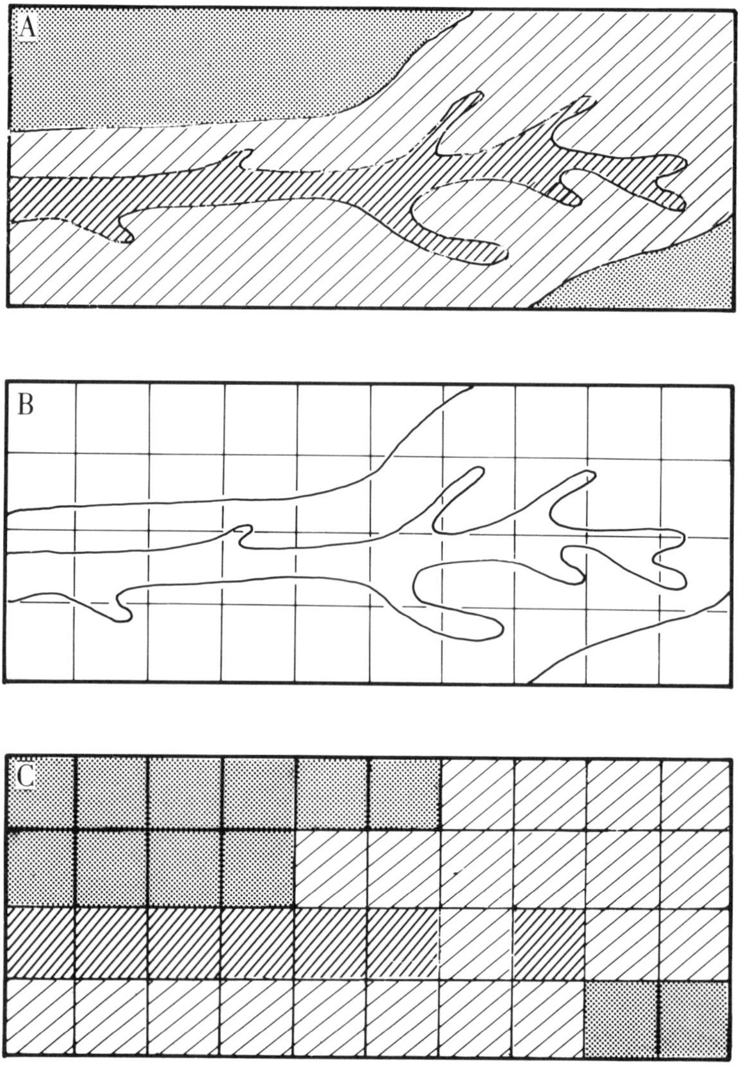

Figure 6.1. Example of soil landscape sampled using uniform cells. Here an idealized landscape pattern (A) has been sampled by superposition of a network of uniform cells (B). Because each cell can be coded only with a single category, the final representation (C) can only approximate the actual areas and patterns in the original pattern. An important consideration when this strategy is used to represent soil patterns in geographic data bases is that soil units are selected on the basis of size and shape rather than on the basis of pedologic characteristics. Therefore, (as an example) alluvial soils that occur in long, thin, delineations may be systematically excluded from such a data base, even though their pedologic characteristics may be of great significance in planning and agricultural considerations.

areas of other states, including Pennsylvania, Georgia, Maryland, West Virginia, and Tennessee, so the conditions described below are not merely local in character, but apply to rather large areas in several eastern states.

The lithology of the Rome formation consists of interbedded shales, sandstones, and dolomites of Cambrian age. Total thickness of the formation varies, but early studies assign thicknesses of 450 to 600 meters (1500 to 2000 ft) (Butts, 1940). Later studies have not assigned a value for total thickness, due to complex faulting and folding. The topography of areas underlain by the Rome is characterized by steep slopes with relatively flat, narrow crests (Figure 6.2). Despite the steep slopes, many of these areas are cleared for pasture or cropland, and form the basis for some of the area's more fertile farmland.

The Rome formation has been subjected to extensive fracturing, shearing, and folding. Exposures are often formed of steeply dipping rocks, so the members are presented to the surface as long, parallel bands of constrasting parent material. The result is an intimate spatial association of three contrasting soils developed from this formation. One mapping unit defined on the Rome has been shown to consist of soils from 4 separate soil orders: Entisols, Inceptisols, Alfisols, and Utisols, occurring in such intimate association that as many as three separate orders have been observed within 7 meters (Campbell and Edmonds, 1984).

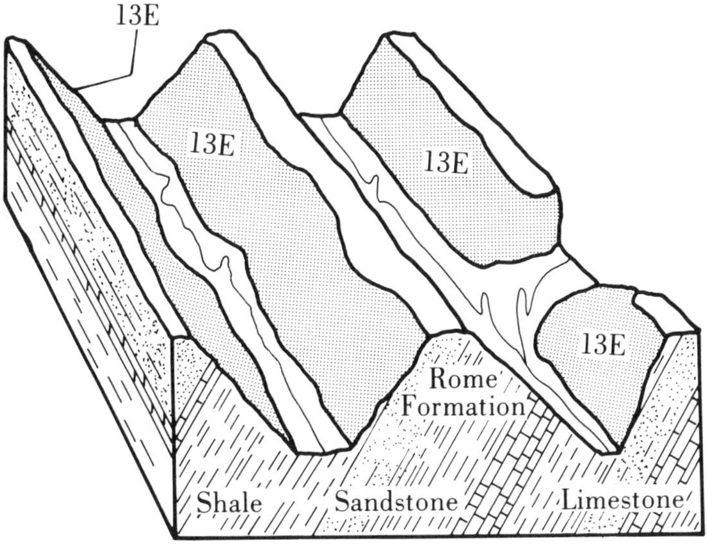

Figure 6.2. Schematic sketch of landscapes formed on the Rome formation, southwestern Virginia. The upturned strata expose contrasting parent material to the surface in a complex spatial pattern that produces an intimate association of highly contrasting soils. Mapping unit 13E includes the several contrasting soils mentioned in the text. (Campbell and Edmonds, 1984.)

The 27 randomly positioned samples positioned within this mapping unit reported by Edmonds and Campbell (1984) are classified into 14 separate families within these 4 soil orders–clear evidence of the range and contrast in diversity of soils developed on this parent material. In the context of the high contrast encountered within this unit, often within very short distances, it is difficult to see how such a mapping unit could be re-defined to isolate the diversity into more homogeneous delineations. The problem is not one of establishment of a broadly defined mapping unit, but rather one of extreme variability within short distances–a natural pattern of a sort that is likely to confound even the most intensive mapping strategies.

The conceptual problem presented by this situation is that mapping units based upon *Soil Taxonomy* express a model of soil variation that represents only one of the many possible kinds of soil units. It focuses exclusively upon those units that can be defined as homogeneous, discrete, soil bodies that are large enough to be delineated at the usual mapping scales. Such a model may be adequate in relatively simple landscapes, but in more complex landscapes, the model is inadequate, and (if one persists in relying exclusively upon it as a mapping device) it becomes necessary to rely upon increasingly severe exceptions from the model to accommodate reality. This is the reason for the elaborate set of different kinds of mapping units defined to implement *Soil Taxonomy* in a mapping context (see Chapter 7). The solution, to the extent that there is a solution, is to accept models for soil variation that recognize the inevitable presence of complex spatial patterns, and the existence of a landscape context for all soil units. Scientists in several nations have established several kinds of soil units that can accommodate the kinds of complex variation encountered in the Rome formation; these units are described in Chapter 7.

7

Taxonomic, Genetic, and Mapping Units

Until now we have focused upon subdivisions of the land surface primarily as abstract entities, without detailed consideration of their meaning. This chapter outlines different kinds of areal and classificational entities used to subdivide the landscape. Although the task of defining landscape units seems at first to be a straightforward process, there are in fact a variety of separate approaches, each with its own logic, and each distinguished from others by basic distinctions in definition.

There are at least three forms of landscape entities: taxonomic units, genetic units, and mapping units. Each subdivides the landscape according to a distinct strategy for representing landscape variation, and each has inherent, significant, qualities for all who study the landscape. Although all three are closely related, and may sometimes only have subtle distinctions, each is based upon a separate logic and serves separate purposes. In some instances (e.g., the "soil series"), the same name designates more than one kind of unit, even though in practice only rarely–if ever–can two or more units exactly correspond to one another. As a result, it is especially important to establish the essential distinctions between these key concepts.

TAXONOMIC UNITS

Taxonomic units form the classes within a classification system (Figure 7.1); for the classification established by *Soil Taxonomy,* these units are the specific orders, suborders, great groups, subgroups, families, and series that compose the system for classifying soils. Taxonomic units are conceptual, abstract entities that exist independently of the spatial domain. There is no constraint upon the geographic manifestation of a specific taxon, which can express itself (for example) as a single large areal unit, or as a multitude of tiny areas; *Soil Taxonomy,* or any taxonomic system, is indifferent to such distinctions. The spatial manifestation of a taxon is therefore completely independent of its definition. From a geographic perspective, taxonomic units are created by the application of taxonomic criteria to the landscape; the informational boundaries in the taxonomic system need not always correspond to observed discontinuities on the landscape.

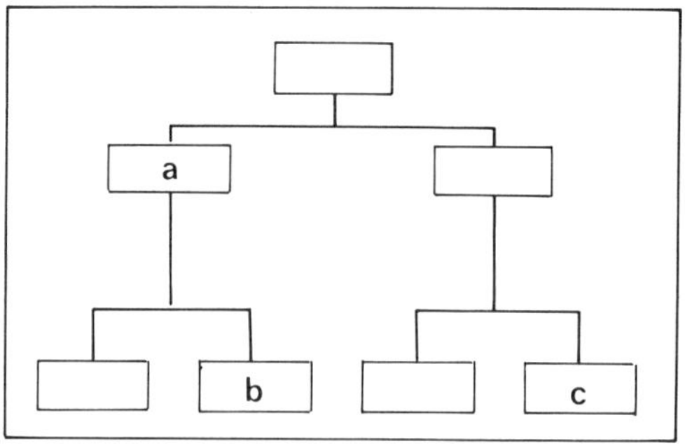

Figure 7.1. Taxonomic units. a, b, and c represent separate taxonomic units.

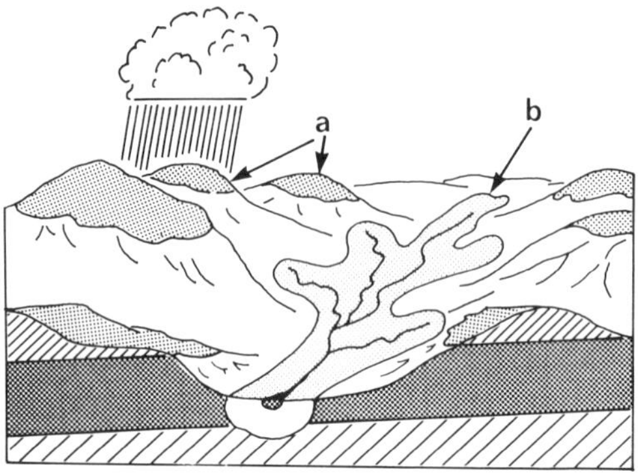

Figure 7.2. Genetic units. Here a and b represent different genetic units.

GENETIC UNITS

Genetic units exist as separate areas within the landscape that have been subjected to the essentially uniform action of the soil forming processes over time (Figure 7.2). That is, at some level of detail, they can be considered to be uniform in respect to origin. Genetic units therefore exist purely in the spatial, or geographic, domain; each possesses size, shape, depth, topographic position, neighboring units, variability, relief, etc., and cannot be isolated from these characteristics (Hole, 1953). They *must* be considered as areal units; to examine individual profiles, without considering relationships to the remainder of the unit, is to ignore the essential features of the unit. The borders of genetic units are not constrained to match taxonomic boundaries. Taxonomic boundaries may in fact split ostensibly homogeneous landscape units–evidence of the fundamental differences between the two kinds of units.

Internal variation within a genetic unit may include several taxa if specific genetic processes produce varied morphology within short distances. For example, frost heaving might produce a single unit characterised by varied microrelief, leading to varied profiles within the same genetic unit. Or contrasting parent materials that have been folded or faulted to produce complex spatial patterns within small areas can produce genetic units with contrasting properties.

MAPPING UNITS

Mapping units, like taxonomic units, are the creation of mankind, artifacts of efforts to represent soil distributions on maps. Mapping units correspond to legend items on a map; they are subdivisions defined for the purpose of depicting on maps the landscape units observed in the field (Figure 7.3). Because limitations imposed by scale and legibility do not permit exact representation of all detail observed (or inferred) in the field, mapping units cannot correspond exactly to the taxonomic or genetic units that they represent. At some level of detail, mapping units are often assumed (by the conceptual models employed by most maps) to be homogenous, and are usually represented cartographically as uniform entities.

In theory, a mapping unit could correspond exactly to a taxonomic or genetic unit; one might represent, using a single symbol, the distribution of all Paleudults within a given area. In practice, however, the limitations of our information, and of cartographic detail and legibility prevent establishment of a one-to-one correspondence between mapping units and the taxonomic or genetic units they represent. Mapping units therefore typically include areas of soils other than those specified in the legend.

Mapping units named for one taxon may encompass more than one taxon; the inclusions of unnamed soils may be too small to map, or occur in a pattern too complex to specify. An important practical problem in most national soil surveys is that the map user has no way to learn of the amount, character, and pattern of these variations, because he or she must rely completely upon the map, its legend, and the accompanying report, which usually presents little detail on such matters. The typical mapping unit includes an unknown amount of variation, usually said to be insignificant in respect to management of the soil.

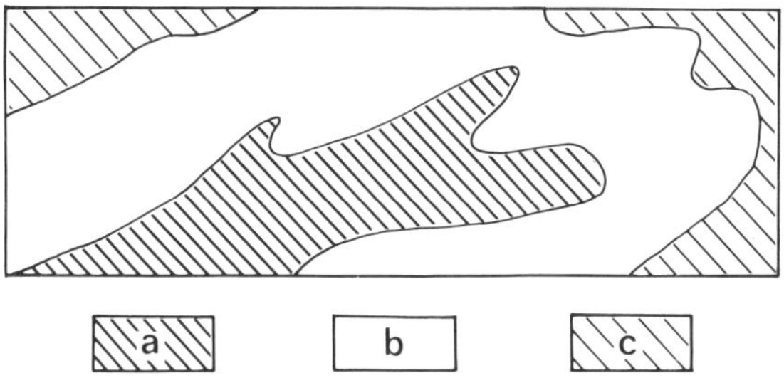

Figure 7.3. Mapping units. a, b, and c designate separate mapping units.

SPECIFIC PEDOLOGICAL UNITS

Scientists throughout the world have defined a variety of different kinds of pedological units as a means of conveying information about the soil landscape by means of maps and written descriptions. The following sections briefly describe some of the most important of these units, as an introduction to the variety of approaches that have been devised for the study of soil geography. Note that some kinds of units are clearly recognizable as genetic, mapping, or taxonomic units, whereas others possess features of two or more of these kinds of units. Note also that a single name (most notably the soil series) has been applied to refer to different kinds of units. These descriptions are intended only as capsule definitions; full descriptions, as presented in the original references, require lengthy and elaborate explanation.

The Soil Series. (Soil Survey Staff, 1975). One suspects that pedology inherited from geology the practice of identifying very specific landscape entities, searching for their origin, naming them after local features, and correlating them from one occurrence to another. The soil series was officially defined by the U.S. Bureau of Soils in 1903 to designate collections of soil landscape units having similar origin but differing texture. (Individual series had been defined prior to this date.)

> "The series is a group of soils having the same genetic horizons, similar in their important characteristics and arrangement in the soil profile, and having the same parent material. Thus the series comprises soils having essentially the same color, structure, natural drainage conditions, and other important internal characteristics, and the same range in relief. The texture of the soil profile, especially in the plow layer, may vary." (Ulrich et al, 1947, p. 12)

In its original meaning, the soil series was a collection of soils that varied in respect to texture of the upper portion of the soil profile. The soil series, in this meaning, is a plural term that identified a group of closely related soils.

In current usage "soil series" has a much more narrowly defined meaning that for most scientists carries a singular rather than a plural meaning. The soil type is no longer recognized as a subdivision of the series, and the definitions of individual series have been narrowed significantly from those originally defined in the early 1900s (Simonson, 1968). For example, the Miami series, as orginally defined and correlated, occurred in states from New York to North Dakota, and as far south as Mississippi, and included soils now placed in three separate soil orders. As more information concerning origin, characteristics, and behavior of soils was acquired over the years, such broad definitions were redefined to produce much narrower classes. For example, in Indiana alone, (see Figure 3.1) some 13 separate series (in addition to the present Miami) have been created since 1906 from the original series established in 1900 (Sanders et al, 1979). Because of refinements of this kind, as well as establishment of new series in previously unmapped areas, the number of soil series officially recognized in the United States now exceeds 10,000.

Soil Taxonomy places two major constraints upon definitions of the soil series. First, a series cannot be defined if its creation bridges the boundary between two higher taxa in the taxonomy. Second, similar series must be distinguished by restrictions upon at least one property that defines the parent soil family. (This requirement assures that the system will be hierarchical; each series must belong to a family. The criteria that separate series must, by definition, define a range that exceeds that of measurement errors for the properties in question.)

Recognized variations within a mapping unit based upon a soil series are not permitted to reflect differences in soil genesis, except to the extent that they lead to soil or landscape qualities that influence use and management. Such variations (perhaps in slope, or stoniness, for example) may be mapped separately as *phases* of a series. Of course, variations in genesis *do* exist within areas mapped as a single series, although they are not explicitly recognized. Some of these variations are simply natural fluctuations in soil properties that occur at a scale too fine to map. These variations are seldom explicitly recognized although most users of soil surveys implicitly accept their presence. Other variations are caused by inclusions of soils other than the named series. *Soil Taxonomy* permits a mapping unit based upon a soil series to include a single contrasting soil series as an inclusion if it does not exceed 10% of the area of the mapping unit, and if delineations occur in parcels too small to represent at the scale of the survey. Inclusions of pedons differing from the named series are permitted up to 50% of the area of the mapping unit, if the inclusions are similar to the named series (Soil Survey Staff, 1975a, pp. 408–409).

The concept of the soil series provides a vehicle for defining identifiable, distinct, soil units. It forms the basis for concepts such as the soil association and the soil complex. The practical significance of the soil series as a concept can be measured by its acceptance by other national soil surveys, which are presumably free to employ whatever devices they judge to be most effective for their purposes. The soil surveys of England and Wales, France, and Japan, for example, all employ the term "soil series" to define large scale mapping units. Although details vary from one nation to the next, all appear to use the term "soil series" in a manner analogous to its definition in the United States. Boulaine (1980) may have identified one of the most attractive practical features of the soil series when he stated that the soil series provides both specific and general information about a specific soil. A person familiar with a specific series has at once very specific information about a certain profile class, as well as more general information implicit in knowledge of its placement within the larger taxonomy.

The Pedon. (Soil Survey Staff, 1975a) The pedon is said to be the basic spatial entity of *Soil Taxonomy.* It appears to be an original concept first expressed in *Soil Taxonomy* and its predecessors. (Jenny's [1965] "tesserra" appears to be an analogous idea, but emphasized the actual soil volume sampled, which is usually much smaller than a pedon.) The pedon is defined as a minimum sampling unit that encompasses 1 to 10 square meters of surface area, depending upon local variability. In depth, it extends downward to the interface between genetic horizons and geologic material, or to the depth of deepest plant roots, whichever is deeper. The lateral dimensions can extend, if necessary, to encompass one half the wavelength of cyclic or intermittent horizons. Its boundaries all form contacts with other pedons or non-soil materials.

The pedon is an arbitrary unit (in Knox's (1965) meaning) that forms a conceptual foundation for the study of soils as geographic entities. The institutionalization of this concept marks an important milestone in the development of studies of soil classification and soil variation, but the current definition is susceptible to criticism from several perspectives:

(a) Webster (1968) notes that the pedon is "irrelevant" to *Soil Taxonomy*. The term "pedon" is used throughout *Soil Taxonomy*, but the *concept* is not. The de facto focus is upon the soil profile, not the pedon.

(b) It is difficult to know how to interpret many elements of the definition. For example, the rates of variation of separate properties over distance are quite different (Campbell, 1977). Unless we are to place arbitrary emphasis upon certain properties (something not permitted by *Soil Taxonomy*), we have no basis for deciding which properties to use as the basis for defining the limits of the pedon.

(c) The definition appears impractical to implement in the field by using the objective, quantitative criteria specified by *Soil Taxonomy* (Edmonds, 1983). Literal application of the definition as presented by *Soil Taxonomy* (Soil Survey Staff, 1975a, p. 3) requires assessment of the spatial periodity of the numerous properties required for complete characterization of a pedon for classification when the place to place variation of horizons is suspected to be intermittent or cyclic. As a practical matter, acquisition of sufficient information for this determination would seem to require such intensive sampling that the pedon under study would by destroyed by the process of determining its extent.

The Polypedon. (Soil Survey Staff, 1975) The polypedon is a taxonomic landscape unit consisting of two or more contiguous pedons with properties that meet the definition of a single soil series. A polypedon adjoins non-soil entities on all sides, or pedons and polypedons assigned to other soil series. Ostensibly the polypedon is a geographic entity, but in fact the definition in *Soil Taxonomy* establishes it as a taxonomic unit.

The polypedon is defined as a taxonomically homogeneous unit; all pedons within a polypedon must belong to a single soil series. The range of variation permitted within a polypedon is controlled by the definition of the parent soil series, but *Soil Taxonomy* (p. 5) prohibits assignment of differing taxa to the same polypedon. The polypedon is therefore a taxonomic unit.

In nature, it is possible for a soil landscape unit to be composed of pedons belonging to more than one taxon, either because of complex geologic and ecologic patterns, or because properties within a soil landcape unit fall at the boundary between taxa, causing minor variations within the unit to be placed into differing categories. Under these circumstances, strict application of taxonomic criteria produces either of two results—both seem illogical from a geographic perspective.

If the differing pedons are contiguous, the taxonomic boundary would divide the landscape unit into two sets of pedons, each belonging to a separate taxon. The landscape unit is therefore divided arbitrarily into two segments on the basis of a classification decision isolated from specific details of this particular landcape unit.

If, on the other hand, the differing pedons within this landscape unit are thoroughly intermingled, it must then be segmented into a complex pattern of many small isolated pedons. (Pedons from different taxa cannot be placed in the same polypedon.) The landscape unit, in this instance defined by the complex pattern of contrasting soils, is therefore dissected into its component elements. Its basic geographic properties are fragmented. In a mapping context *Soil Taxonomy* circumvents these problems by defining several kinds of mapping units, as described in the following sections.

Consociations. (Soil Survey Staff, 1975b) Consociations are mapping units composed mainly of phases of single taxon at the type and phase levels of classification which are not included in *Soil Taxonomy* (Soil Survey Staff, 1975a). At least one half of the pedons within a given consociation belong to the unit (usually a soil series) that names the mapping unit. Dissimilar inclusions are permitted to a maximum of 15% of the area if they form a limiting factor for use and management, or up to a maximum of 25% if they do not limit use an management. A single contrasting inclusion should not exceed 10% of the areal extent. Three-fourths or more of the polypedons of the unit fit within the taxon that provides the name for the mapping unit, or within defintions of similar soils.

Complexes. (Soil Survey Staff, 1975b) Complexes are mapping units formed by two or more contrasting taxa occurring in a recurring spatial pattern so complex it cannot be cartographically resolved at a scale of 1:24,000. The contrast between member soils is so great that definition of a consociation is not appropriate. The term has also been used for assemblages of bodies of contrasting units occurring in patterns that change so frequently (on alluvial flood plains, for example) that detailed mapping of them would be of no practical value.

Associations. (Soil Survey Staff, 1975b) Soil associations are mapping units similar in definition to that of the soil complex, except that the geographic pattern is coarse enough to be resolved at 1:24,000. Associations are usually defined for use on small scale generalized maps, including the county soil association maps used in soil surveys in the U.S. *Undifferentiated groups* (Soil Survey Staff, 1975b) designate mapping units composed of taxa arranged in consistent geographic patterns; contrasts between member taxa differ so little, or are so changeable in respect to use and management from season to season (as on alluvial flood plains), that there is little justification for forming separate mapping units from its components.

Taxajuncts. (Soil Survey Staff, 1975b) Taxadjuncts consist of polypedons with characteristics outside the range of any recognized soil series, but with use and mangement interpretations differing only slightly from those of established series. As a result, a taxadjunct is treated as a member of the most similar established series for use and management interpretations, although observed differences between the taxadjunct and the established series are specified.

Some of the terms just mentioned are foreign to many who may not have everyday access to the practices and policies of the U.S. Soil Conservation Service, since they have no analogues in corresponding classification systems of other national soil surveys, or in the earlier practices of the S.C.S. Other terms may be more familiar. For example, the soil complex, and the soil association have been used for many years in the context of small scale mapping, where requirements for legibility require the establishment of generalized mapping units in areas of complex pedological patterns. For many scientists, the use of these kinds of mapping units in the context of large scale (i.e., 1:24,000 and larger) mapping may seem unfamiliar. These kinds of units are required, it seems, because of the uneasy relationship between the taxonomic units defined by *Soil Taxonomy* (Soil Survey Staff, 1975) and the natural occurrence of landscape units on the terrain. Literal application of the requirements of *Soil Taxonomy* to landscapes of even modest complexity can result in a logically and cartographically complex pattern of diverse pedological units; ostensibly reasonable mapping units are found, upon closer examination, to be composed of contrasting taxonomic units (Edmonds, 1983). Taxadjuncts, consociations, and some of the other units defined above represent an effort to provide a coherent framework to the logical and cartographic confusion created by the application of *Soil Taxonomy* in a mapping context.

Note that many of the definitions given above specify the nominal characteristics of given mapping units, especially in respect to the percentage composition of units. The user of a soil survey has very little explict knowledge of how these idealized definitions might be applied to specific landscapes, or how such applications might vary from one context to another. Certainly much of the precision and consistency implied in the formal definitions seem suspect, as it seems unlikely that most soil surveyors will have the time or resources to gather much more than rough estimates of the areal extents of inclusions, and it is not clear if the percentages are to apply to the mapping unit as a whole, or to each delineation. Futhermore, the notion of contrasting soils seems likely to vary widely, as use and management varies greatly, sometimes even within relatively small regions.

Elementary Soil Areals and Soil Combinations. (Fridland, 1974) The work of Soviet soil scientists and their predecessors in czarist Russia forms a long and sophisticated, if complex, record of the study of soilscapes. Fridland (1974, 1976a, 1976b) and his colleagues have developed what is probably the most comprehensive, and the most fully documented body of research on the nature of the soil landscape.

The term "soil cover" defines the entire population of soils of a region that together form a physically continuous mantle over the earth's surface, interrupted on occasion by rock outcrops and water bodies. Fridland characterizes the soil cover as having a "discrete-continuous" distribution exhibiting both abrupt and gradual transitions between neighboring soil units. The soil cover is composed of elementary soil areals (ESAs), defined as soils belonging to a single taxon of the lowest rank, occupying a space that is bounded on all sides by other ESAs or non-soil formations.

The exact distinction in definition between an ESA and the areas of sometimes greatly differing soils within ESAs is not immediately obvious. Study of examples cited by Fridland (1974, 1976a) leads to the following interpretation of the concept. The "limiting" or "initial pattern elements" that may be within ESAs are very small in area–typically only a few square meters–so, the areal extent of the unit is apparently significant. Second, and more significant, is that the genesis of initial pattern elements can be attributed to the effect of a specific organism, or a single aggregation of organisms (e.g., the pedological result of a single tree-tip mound produces a "limiting pattern element," whereas the pedological result of a forest floor consisting entirely of tree-tip mounds and intervening hollows would form a collection of such soil areas that together constitute an ESA. Thus, the single, highly localized effect of a specific tree, a specific anthill, animal burrow, etc. does not constitute an ESA. To use U.S. terminology such effects might lead to the genesis of individual pedons.

As a result, not all ESAs are uniform (Figure 7.4). Homogeneous ESAs are apparently definitionally akin to polypedons, except that each ESA is judged individually on its geographic properties, including slope and other geographic qualities outside the concept of the soil series (as defined in the U.S.). All profiles within a homogenous ESA must belong to the same taxon at the lowest level in the classification system. Boundaries must border other ESAs or "not soil." Heterogeneous ESAs on the other hand include at least two separate taxa. Sporadically patchy ESAs are formed by irregular occurrences of small bodies of slightly (or, if contrasting, of biological, and hence ephemeral in nature) differing taxa set in the larger body of the predominant taxon. The other form of heterogeneous ESA is the regular-cyclic ESA, exemplified by a continuous network of polygons, or the hexagonal cracking of expanding clay soils.

ESAs occur in soil combinations–consistent spatial arrangements of ESAs caused by pedogenetic processes; soil combinations are not solely cartographic units, but are the natural genetic fabric of the soilscape. Soil combinations are elementary cells that repeat over distance to form the spatial structure of the soil mantle. The genetic links between ESAs may be formed

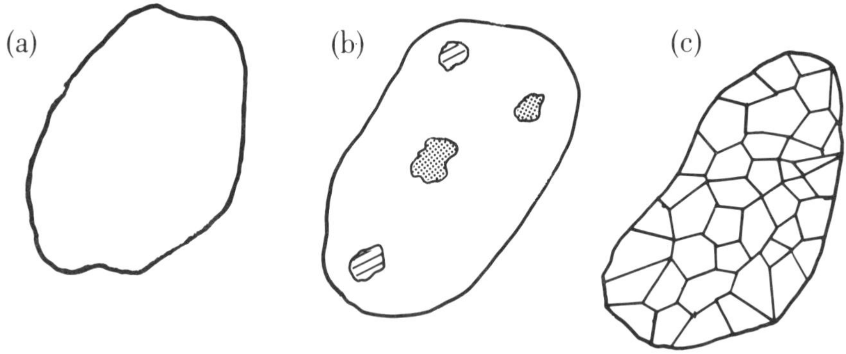

Figure 7.4. Soil units proposed by Fridland include (a) homogeneous elementary soil areal, (b) sporadically patchy elementary soil areal, and (c) regular-cyclic elementary soil areal. (Based upon Fridland, 1976a; Campbell and Edmonds 1984.)

by the transport of materials in two directions ("bi-lateral links", or duo-flow, using the terminolgy from Chapter 3) in a single direction ("uni-lateral links", or mono-flow), or may be poorly expressed or absent. Fridland classifies soil combinations based upon the nature of links between member ESAs and the degree of contrast between them.

The Génon. (Boulaine, 1969) Boulaine (1969, 1980) accepts the notion of the pedon, but bases his ideas upon a definition slightly modified from that of the USDA: an "elementary volume necessary and sufficient to define, at a given instant, the entirety of the structural and material components of the soil" (Boulaine, 1969, p. 35, as translated). He defines a number of related concepts, including the *pédode* (the length of time necessary to examine the dynamic features of a pedon, the *pédôme,* the series of conditions experienced during the pedode), and the *péripédon* (the volume of the pedosphere surrounding the pedon and influencing its character). He recognizes the soil series as an important feature of our descriptions of the landscape: "the soil series is [established] at a level so detailed that its definition is in practice independent of the [taxonomic] system to which one contemplates its assignment" (Boulaine, 1980, p. 101, as translated). He therefore appears to accept the common occurrence of natural soil units (in the meaning of Van Wambeke, 1966) as fundamental components of the soil landscape.

The central concept in Boulaine's work is the *génon,* a mapping unit applicable at scales of 1:20,000 to 1:100,000. The *génon* defines a volume of soil "including all of the pedons possessing the same structure, the same characteristics, and resulting from the same pedogenesis" (Boulaine, 1980, p.102). "In a *génon,* there is a unity of pedogenesis, a coherent organization . . ." (p. 103). *Génons* composed entirely of pedons belonging to a single taxa ("monotaxons") include *génons simples* (collections of pedons each 1 meter square or less), and *génons variants* (collections of pedons each 10 square meters in area, or more). *Génons simples* are either *homogènes* or *mâtinés* (lit.

Plate 7.1 Upturned beds of contrasting geologic strata observed in a road cut near the Virginia–West Virginia border west of Narrows, Virginia. Because the surface expression of this formation forms parent material for development of soils at the land surface, this example illustrates well the factors that lead to the complex composite soil patterns described in the text.

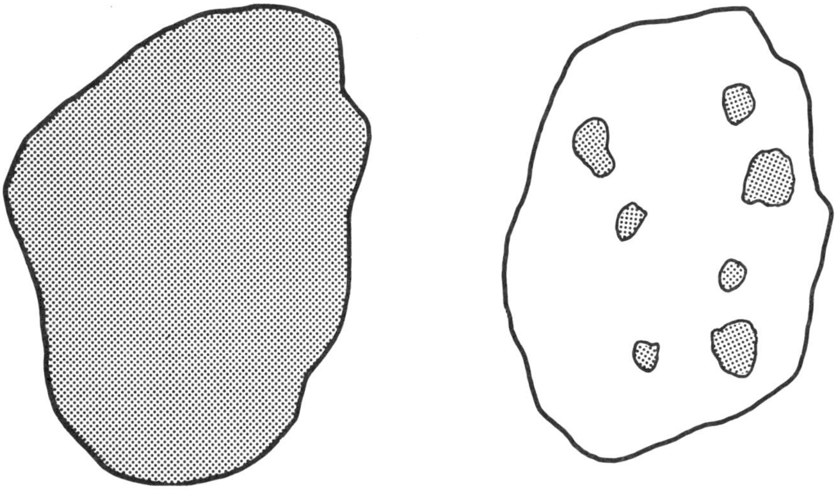

Figure 7.5. Examples of the soil units proposed by Boulaine. Shown here are the *génon simple homogène (left)*, composed of a single soil, and the *génon simple mâtiné (right)*, composed of at least two kinds of soil, symbolized here by the shaded pattern that contrasts with the more extensive background. Here the included soils do not occur in a characteristic arrangement. (Based upon Boulaine, 1969, 1980; Campbell and Edmonds, 1984.)

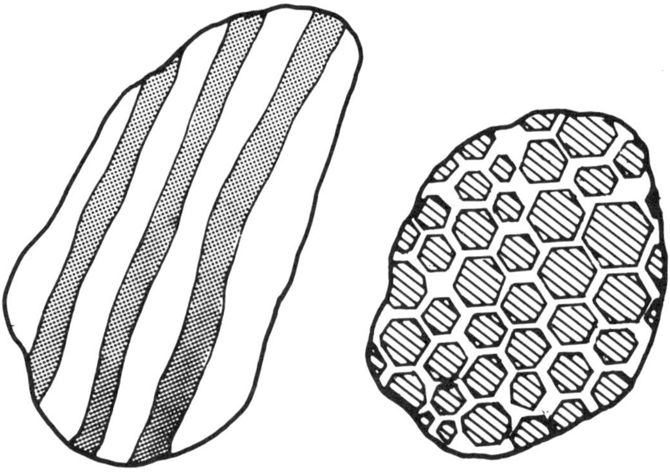

Figure 7.6. Additional units proposed by Boulaine, including the *génon variant strié* (left), and the *génon variant cellulaire.* In both units, the soils that contrast with the more extensive background occur in distinctive patterns depicted here in rather stylized representations. Designation of a unit as a *génon variant strié* (for example) informs the reader of the presence of contrasting soils within the mapping unit, and the general nature of their occurence within each delineation, even though the specific patterns are not represented. (Based upon Boulaine, 1969, 1980; Campbell and Edmonds, 1984.)

"mongrel," or "cross-breed") (Figure 7.5). *Génons variants* are designated by the geographical arrangement of member pedons: e.g., *génon variant strié* (stripes or bands) (Plate 7.1), *génon variant cellulaire* (cell-like patterns, analogous to the patterned ground of periglacial landscapes), and *génon en moasïque* (mosaic) (Figure 7.6). Another broad class of *génons* are composed of pedons belonging to two or more taxa–*génons complexes.* Boulaine characterizes *génons complexes* on the basis of internal contrast, character of member pedons, and relationships between member soils. Individual *génons complexes* can be described by the geographic and geometric arrangement of member pedons. Boulaine (1980) gives examples of several forms of *génons complexes* (e.g., *génons complex orientes* and *génons complexes cellulaire.*) Boulaine's ideas are presented here in only their briefest form, yet it should be clear that his approach considers entire units of a landscape in their spatial and pedological settings, not just arbitrary units isolated from their context in the landscape.

Application of Boulaine's approach need not require collection of more detailed information, or even the application of more detailed cartography, than that presently used in many soil surveys. Rather it would seem sufficient to represent much of information now known to soil surveyors regarding the nature of mapping units, using mapping units defined in the style proposed by Boulaine. Designation of a unit as a *génon strié* need not require the actual mapping of the pattern within each delineation, but simply the application of the knowledge, already possessed by the soil surveyor, that the unit is characterized by a specific pattern of variation. Thus, his approach would appear to form the basis for an economically feasible strategy for mapping and symbolizing soil patterns.

Boulaine's overall approach is consistent with Lyford's statement advocating efforts ". . . to give the average map user a greater awareness than he now has of the complexity that does exist in many mapping units" (1974, p. 207), and of Wilding and Miller's call for ". . . more *not* less comprehensive description of mapping units including *kinds, amounts* and *spatial distribution* of component soils" (1979, p. 92, their emphasis).

The Pedotop. (Hasse, 1968). This definition of a geographic unit of soil is in some respects analogous to the pedon and polypedon, but differs from the pedon with respect to its greater range of variability, and from the polypedon in respect to size and continuity. The *Pedotop* is spatially closed, is geographically homogeneous according to its pedological attributes (Hasse, 1968, p. 76), and is created by a uniform combination of soil forming factors. Yet the *Pedotop* is by no means completely uniform (Figure 7.7).

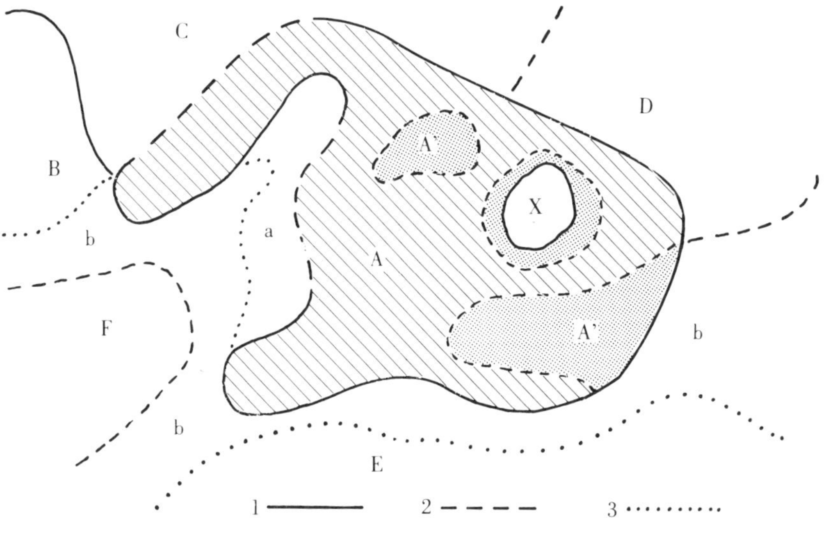

Figure 7.7 The *Pedotop,* as defined by Hasse, 1968. This diagram represents, in schematic form, "Pedotop A." A, B, C, D, E, F, and X represent separate genetic units (polypodons). a designates a transitional zone from A to adjoining landforms; A′ denotes a variation of A; b designates a transitional zone from adjacent landforms; 1, 2, and 3 represent variations in distinctiveness of boundaries in sequence from sharp to progressive. (Based upon Hasse, 1968, as modified by Campbell and Edmonds, 1984).

Its definition explicitly recognizes (a) zones of transition (of varying width) to neighboring units, (b) variation from the modal concept within the unit, and (c) inclusions of dissimilar soils, both from adjacent soils, and from others not adjacent. An overall characteristic of the *Pedotop* is its uniform ecologic structure; this does not, however, imply a uniform composition, since a structured pattern of variation is included within the definition.

In the writings of Hasse we can see once again, as with those of Fridland and Boulaine, an emphasis upon examination of terrain units from a broad spatial and pedological perspective, one that contrasts greatly with the rather narrowly defined view exemplified by the pedon and the polypedon.

8

Generalizations of Soil Maps

Generalized maps form a significant means of representing soil distributions, and therefore deserve the attention of all who make and use soil maps (Bartelli, 1966). This chapter examines processes used in generalizing soil maps, identifies the kinds of errors they produce, and outlines some of the broader issues of significance in generalization.

Generalizations are required to reduce complex data into a form suitable for comprehension of their essential elements. Generalizations of tabulated data are common in many contexts and, as a result, are easily understood both in concept and in practice (Table 8.1). In contrast, generalizations of spatial data, as presented on maps, although no less common, are much more complex both in theory and in implementation. As a result, even the most widely used methods of cartographic generalization often are not clearly understood by makers and users of generalized maps.

GENERALIZED MAPS AND SCHEMATIC MAPS

If information presented in detail on one map is compiled in less detailed form on another, then the second map is said to be a generalized version of the first. Generalized maps are therefore based upon logical and/or visual re-organizations of information from more detailed maps, rather than upon original field surveys (Varnes, 1974). This distinction is simple, but fundamental to any discussion of generalization; original field surveys can be compiled at varied levels of detail, and can therefore display the coarse spatial detail and small scale that we often associate with generalized maps, even though such maps are not true generalizations.

Typically, generalized maps are designed to depict a broad regional overview, and as a result, are presented at smaller scales than are the original source maps. In contrast, schematic maps, which share with generalized maps the characteristics of coarse detail and broadly defined categories, are approximations of a pattern based upon reconnaissance surveys, map analysis, or other estimates of a pedologic pattern (Orvedal et al, 1949; Simonson, 1971). A schematic soil map might be prepared from information presented on geologic, climatic, and vegetation maps, whereas a generalized map is based upon a soil map prepared in detail from a field survey. The generalized map is therefore a simplification of a distribution known in

Table 8.1 Generalization of Tabular Information by Aggregation at Selected Taxonomic Levels

Order	Subgroup	Series	
Ultisols (138,340)	Typic Hapludults (104,590)	Appling	(75,000)
		Cecil	(4,300)
		Durham	(700)
		Masada	(5900)
		Mayodan	(18,600)
	Typic Fragiudults (13,750)	Bourne	(11,850)
		Vacluse	(1,900)
	Typic Paleudults (20,000)	Norfolk	(3,300)
		Lucy	(5,500)
		Orangeburg	(11,200)
Alfisols (4,235)	Ultic Hapuldalfs (3,325)	Enon	(400)
		Pamunkey	(2,925)
	Typic Ochraqualfs (1,000)	Forestdale	(1,000)
Entisols (650)	Typic Udifluvents	Tocca	(400)
	Typic Udipsamments	Buncombe	(250)
Inceptisols (700)	Dystrocrepts (700)	Louisburg	(700)

Note: Values in parentheses give areal extent in acres.

Source: Hodges et al. (1978).

detail; the schematic map is an estimate of an unknown pattern yet to be mapped in detail.

GENERALIZED MAPS AND DETAILED MAPS

It must be recognized explicitly that all maps form generalizations of reality, and, as a result, a broad interpretation of the term "generalized map" might include all maps, except those published at 1:1 (see Hakanson, 1978, for an account of such a map). Also, even most detailed maps are, of course, generalized versions of field sheets and manuscript maps used to compile published maps. The term "generalized map" is used here, in recognition of these facts, to designate the results of generalization processess applied to a given map. For example, we can generalize the map pattern represented on a published soil survey, even though we recognize that the published map is itself a generalization of the field sheets used to produce the survey. Thus, the map that is to be generalized may be an original field survey, or it may itself be a generalization of another, more detailed, map. In the latter instance, the source map is a "detailed map" only in the sense that it is detailed relative to the generalization to be made.

SIGNIFICANCE AND APPLICATIONS

Generalizations are necessary because of the requirement to examine distributions at varied levels of detail. Whenever the scale of consideration focuses upon larger areas, detail must be reduced to present information at scales compatible with that of the interest of the map user (and compatible as well with graphic legibility and the ability of readers to perceive detail). Therefore, to the extent that maps are actually used in the decision-making process, generalized maps are used in forming regional and/or national policies (as opposed to strictly local site-specific decisions). Because generalized maps may be used in zoning, planning, legislative, and other legal and quasi-legal contexts, deficiencies in generalization processes may have subtle, but far-reaching, effects.

Although generalization is especially important for maps used by laymen, small-scale generalized maps must also be used by specialists. Generalized maps form important tools for scientific studies because the recognition of broad-scale patterns, and of interrelationships between regional environmental patterns, requires examination of information distributed over large areas. For example, Hathaway et al (1979, 1980) found generalized soil maps of western Kansas to be useful indicators of the chemical quality of water from irrigation wells.

Most people can recognize from their own experience the difficulties encountered in the perception and assimilation of fine cartographic detail over large map areas. For example, the soil survey of Chesterfield County, Virginia (Hodges et al, 1978) includes 134 mapping units and some 10,000 individual delineations. Examination of even one of the 63 individual map sheets is a challenge to both the eye and the mind, and assimilation of information for the entire county in its most detailed form would clearly be impossible. Simonson (1971, p. 963) discusses this problem, and suggests that county soil association maps include no more than 18 mapping units. "It is easy for [a soil scientist] to place more information into a map than its users will need . . . Need for a ceiling on numbers of mapping units follows from limitations of the human mind, from a ceiling on what it can comprehend at one time."

In addition, generalization is required to render map detail suitable for reproduction at small scale. Because mechanical limitations of drafting, photographic, and printing processes often prevent legible reproduction at small scale, generalizations are required to create simpler, less detailed, map patterns. Generalization procedures include smoothing of lines, deletion of delineations and boundary segments, and combinations of detailed categories into fewer, more broadly defined categories.

Although generalization achieves simplicity at the expense of decreased accuracy or precision, and thereby creates "error" in respect to the detailed map, such errors are in fact inherent to all maps, and are accepted by experienced map users. In its broadest sense, generalization is unavoidable; the challenge is to choose a generalization strategy appropriate to the purpose of the map and the pattern to be represented. Therefore, a knowledge of alternative procedures for cartographic generalization is essential preparation for efficient compilation and use of generalized maps.

THE SOIL MAP AS A REPRESENTATION OF THE LANDSCAPE

Effective generalization requires knowledge of the pattern to be generalized, and of the cartographic model used to represent the pattern.

The usual soil map belongs to a family of maps in which a distribution is represented as a mosaic of discrete parcels, or delineations. Each parcel is assigned to one of several mapping units, usually symbolized by colors and/or alphanumeric symbols. Mapping units usually do not correspond directly (either in theory or in practice) to taxonomic units due to unmapped inclusions within delineations and the use of associations and complexes to map intricate patterns. Monkhouse and Wilkinson (1971, p. 38) refer to soil maps as representatives of "chorochromatic maps" because "units do not involve any consideration of quantities or values." Although this statement is true for many maps that have a similar appearance (e.g., land cover maps), it is of course not true for soil maps, as quantities and values are essential considerations in making and interpreting soil maps. Although soil maps have been constructed using the isopleth principle, this discussion considers only the usual pedologic maps based upon delineation of discrete landscape units.

More specifically, this chapter examines generalizations of detailed maps, such as the large scale SCS county soil survey maps in the United States. The kinds of generalizations to be considered are those that selectively retain the same taxonomic system, accept the same system of mapping units, and the same placement of boundaries as do detailed maps. Thus, a generalization is a re-organization of information on the detailed map without remapping the landscape or introducing information not presented by the detailed map. Implicit is that a generalization is usually intended to represent the original distribution at a scale smaller than that of the detailed map.

The boundaries of delineations are, of course, the essence of a soil map. Usually both makers and users of detailed maps focus upon correct placement of boundaries in respect to actual positions of boundaries on the landscape and ignore the varied geographic character of the boundaries. In contrast, generalization processes accept the positions of mapped boundaries but focus upon manipulations that alter the intricacy of boundary form and remove selected boundary segments—operations that require a sophisticated appreciation of the varied meanings of landscape boundaries.

Boundaries between adjacent units are depicted on soil maps as single, continuous, smooth lines of uniform width. In reality, of course, soil boundaries vary greatly in respect to abruptness, intricacy, and contrast between neighboring units (Gile, 1975; Campbell, 1977). In generalizations, such boundary qualities attain a significance not usually portrayed by the usual soil maps. Generalization can proceed only by altering boundaries, and optimum generalization can be achieved only by identifying which boundary segments are to be preferred for retention or exclusion. Because the map itself depicts all boundaries as identical line segments, successful generalization requires identification, from evidence in the map itself, of the hierarchical structure of boundaries. Varnes (1974) and Hole (1978) both present examples of maps that symbolize the hierarchy of boundary values using varied line weights and symbols (Figures 4.22, 4.23, and 8.1). Explicit acknowledgement of this hierarchy permits application of some of the generalization procedures described below.

GENERALIZATION OPERATIONS

Cartographers have examined generalization processes, but have focused almost exclusively upon generalization of individual line segements (Fahey, 1954; Robinson et al, 1978) without examining generalization of areal units and the special problems inherent in generalizations of soil maps. Soil scientists have examined interrelationships between generalization and taxonomy (Orvedal et al, 1949; Buol et al, 1980), but have not systematically examined cartographic generalization processes and the errors they may generate. As a result, there is little explicit reference to alternative generalization processes and errors in the scientific literature of either cartography or pedology. The following discussion of generalization operations must therefore be based

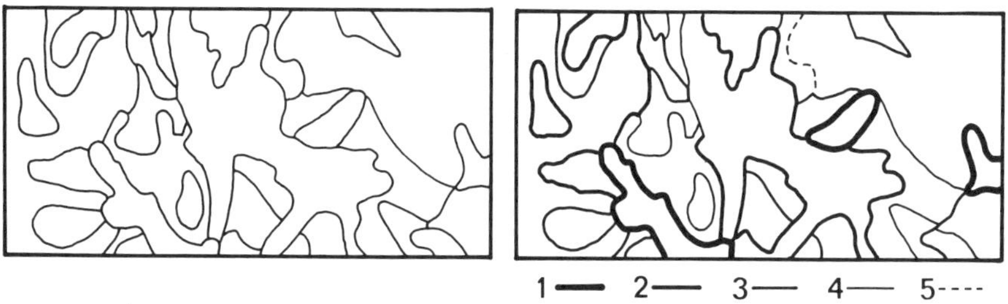

Figure 8.1 Assignment of ranks to soil boundary segments. Left: a portion of the soil pattern in Chesterfield County, Virginia, as depicted by Hodges et al. (1978). Right: the same pattern with ranks assigned to boundary segments according to taxonomic designations of *Soil Taxonomy* (Soil Survey Staff, 1975a). Lines symbolize the lowest contrast between adjacent soils: 1: boundary segments separate different orders, 2: suborders, 3: great groups, 4: subgroups, and 5: families.

largely upon examination of the content of published maps rather than upon explicitly documented procedures.

Smoothing. Boundary smoothing removes intricate loops and indentations that add to the visual complexity of a map. The overall pattern is simplified at the expense of errors created by the transfer of areas at edges of parcels from one category to another. Boundaries on many generalized maps have no doubt been smoothed by draftsmen (rather than soil scientists), who probably follow arbitrary, if perhaps widely accepted, notions of acceptable shapes and configurations of soil bodies. (The broader problem of relating the sinuosity of the generalized line to that of the detailed line has been discussed by Beckett, 1977, and Hakanson, 1978.)

Automated smoothing of digitized lines can be achieved using any of a variety of procedures, including those that preferentially retain points that mark significant changes in direction, those that fit a smoothed interpolation of the original line (in the manner of a moving average), and many others (Douglas and Peucker, 1973). At present little is known about the relative accuracies, their relationships to manual generalization procedures, or the errors inherent to the digitization process itself (Jenks, 1981). Apparently there is little evidence that permits assessment of the influence of such procedures upon representations of the soil pattern.

Deletion of Parcels. At small scale some parcels may be judged to occupy insignificant areas and are therefore omitted from the generalized version. Sizes and positions of parcels may play an important role in this decision. For example, if small delineations are isolated against a large, uniform, background, they may be judged to form only a minor element of the overall pattern, and therefore do not appear in the generalization. This kind of generalization has been applied manually, although computer algorithms have been applied in a research context (Webster and Burrough, 1972).

Several authors have recommended minimum sizes for parcels to be represented on published maps. Boulaine (1980) recommends that the smallest delineations that should be represented on a published soil map correspond to an area of 25 sq mm. The smallest map distance separating two parallel lines should be at least 2 mm. These statements are in approximate agreement with those of Fridland (1976a), who specifies that the smallest delineations on a published map should be no smaller than 20 sq mm. The values that Kellogg and Orvedal (1969) recommend (p. 125) as minimum sizes of delineations on published soil maps translate to a map area of about 40 sq mm. In the United States, soil surveyors following SCS guidelines generally map delineations no smaller than 3 acres (1.2 ha) for maps published at 1:20,000, or 1 acre (0.4 ha) for maps at 1:15,840. These values create map delineations smaller than those mentioned above, but in practice the smallest parcels represented on most detailed soil surveys published in the United States exceed these values by substantial margins.

Enlargement of Parcels. A small delineation may be recognized to be pedologically significant despite its small areal extent. Such areas may be enlarged by generalization to assure that they will be legible on the generalized version. The most common instances of this kind of generalization are encountered in the representation of alluvial soils that occur in long, narrow, delineations; often portions of these areas are represented as areas larger than their true relative areal extent in order to maintain legibility of correctly scaled parcels.

Deletion of Boundary Segments. A soil map can be simplified by removing boundary segments separating pedologically similar units, using the relative ranks of boundaries to select those segments favored for retention or deletion. This process may be identical in effect to combination of mapping units as described below.

Combination of Mapping Units. Mapping units can be combined to form fewer, but more broadly defined, units. For example, we may choose to combine those units that are most similar pedologically—a procedure that is identical to deletion of boundary segments as described above. Other strategies for combining mapping units and delineations are described below. In view of the fact that county soil surveys in the United States may have as many as 100 or more mapping units, this operation is essential for most generalizations. Simonson (1971) recommends that county soil association maps show no more than 18 mapping units; some would argue that even this number is too large. (In the U.S. published county soil association maps often use from 5 to 15 mapping units.)

Addition of Boundaries. In some instances, boundaries may be added as a means of simplifying a pattern. A number of small delineations in close proximity to each other might be enclosed by a single, larger, boundary with interior lines deleted. Or, an island of a soil separated only by a narrow band from a larger delineation of the same soil might be represented as a peninsula of the larger delineation by the addition of boundary segments (and the corresponding deletion of others).

Assignment of Ranks to Boundaries. A soil map can be examined to identify relative strengths of boundaries, as described by Varnes (1974) and Hole (1978) (Figures 4.22, 4.23, and 8.1). Establishment of the ranking of boundary values can be a preliminary step to the deletion of boundary segments (as explained above), or simply to symbolization of differing boundary ranks. The state soil maps of Kansas (Bidwell and McBee, 1973) and Virginia (SCS, 1979), for example, represent differing strengths of boundaries by varied line weights, and major pedologic changes by shifts in the hues and/or intensities of colors that symbolize mapping units. Minor differences between soils are depicted by smaller changes in color, or by retaining the same color across a boundary, but changing the symbol for the mapping unit.

Shift in Boundary Position. In some instances long, narrow delineations would be illegible at small scale if boundaries were shown in their correct relative positions. Therefore, if such delineations are to be retained in a generalization, it is necessary to shift boundaries to parallel positions slightly removed from their correct positions. Such changes, of course, result in the enlargements and reductions in sizes of parcels, as mentioned above.

GENERALIZATION STRATEGIES

Accurate generalization requires application of an overall generalization strategy that controls use of the generalization operations mentioned above. For example, simplification can be achieved by elimination of boundary segments, but if boundary segments are removed without the order provided by an overall generalization strategy, simplification is achieved at the cost of producing a chaotic legend and a useless map. Therefore, all generalizations should be conducted by consistent application of an overall strategy that reduces both visual and logical complexity, while preserving in the generalized version as much of the original arrangement and inventory of soils as may be possible. Simplification, of course, always causes a loss of information; the generalization strategies described below (which can be applied singly or in combination) each preserve, and discard, different features of the original pattern.

Graphic Generalization. Graphic generalization is the manipulation of the shapes of line segements, with the objective of reducing visual complexity (Figure 8.2). (This process is refered to as "cartographic generalization" by Buol et al 1980; their term is not used here because it conflicts with different meanings used in other contexts.) The original legend remains intact, or experiences only minor change. The objective of graphic generalization is to reduce the total line length on the map while retaining all, or most, delineations. (As a result, this process could also be referred to as "line generaliza-

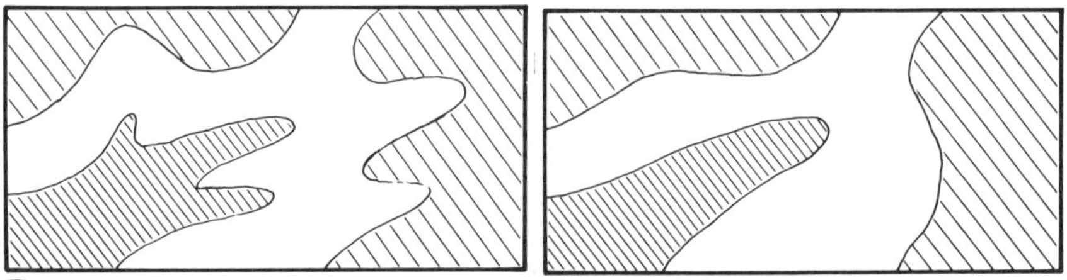

Figure 8.2. Graphic generalization. A detailed map is represented on the left, the corresponding generalization on the right. Boundaries between units retain their identities as defined on the detailed map.

tion.") The soil map is treated as an abstract pattern, with only minor attention devoted to its meaning. For example, a cartographer performing graphic generalization need not understand the character of the two soils separated by the boundary—a distinction that is critical in other forms of generalization described below.

Graphic generalization is applied more or less uniformly to the entire map, with the objective of removing visual complexity. The chief concern is legibility of the final map (which may be published at smaller scale than the detailed map). The kinds of operations necessary to retain legibility include one or more of the following operations mentioned earlier: (a) boundary smoothing, (b) deletion of small delineations, (c) enlargement of small delineations, (d) addition of boundaries, and (e) shifts in boundary positions.

Most generalized soil maps have probably been subjected to graphic generalization. Graphic generalization, if applied independently of other forms of generalization, results in minor changes to the map pattern or map legend. Often it is necessary to apply graphic generalization in combination with other methods to achieve significant simplification of a map pattern.

Taxonomic Generalization. Taxonomic, or categorical, generalization is achieved by removal of boundaries between neighboring parcels that are taxonomically similar (Figure 8.3). The result is a map with fewer, but larger, parcels, and more broadly defined mapping units. Precision is lost because parcels must be relabeled with more general taxonomic identifications, and the legend must be modified to reflect changes in mapping unit definitions. This process does not detract from the *accuracy* of the map (see Chapter 5) but it does reduce the *precision* of information derived from it. (That is, the *identity* of a given area on the ground may be correctly represented, but not at the same level of detail that was possible before generalization.)

Taxonomic generalization can be implemented by either of two strategies. First, parcels can be regarded as characterized by a series of multivariate measurements. The generalization proceeds by finding those categories that display the greatest similarity in respect to these measurements (i.e., those categories that are closest in multivariate data space), then combining these groups into a single composite mapping unit. If these categories share common boundaries on the map, then the boundaries are eliminated and the map is simplified both logically and visually. If the combined categories are not adjacent, then generalization must then consider those categories next in order of similarity in respect to the measured properties. The advantage is that the generalization combines the most similar categories first, assuring that composite mapping units are always formed from the most similar soils. A disadvantage is that legends for individual maps in a series may be quite different because of variations in the pedologic covers of separate regions.

A second strategy for taxonomic generalization is to combine those mapping units that are taxonomically similar. *Soil Taxonomy* (Soil Survey Staff, 1975) lends itself to this form of generalization. For example, soils can be grouped into broad generalized categories at the suborder or great group

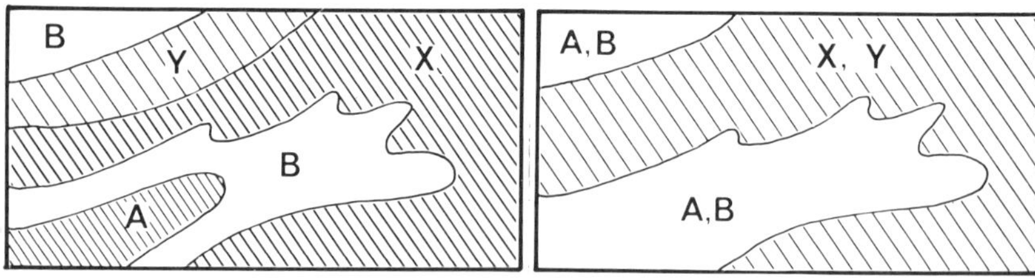

Figure 8.3. Taxonomic generalization combines homogenous mapping units pedologically similar to each other. Generalized map (*right*) combines units A and B and X and Y on the detailed map (*left*). Because pedologically similar soils do not always adjoin one another, taxonomic generalization may not always simplify the map pattern (see parcel B in the upper left corner, which is not simplified by this generalization).

level. Or generalized mapping units could be formed by grouping taxa bearing specific prefixes (e.g., "Aquic-," "Fragi-," etc.). (In theory, this should yield results similar to the first process described above, but in practice the results differ.) Buol et al (1980, p. 346) describe this procedure as the identification and description of mapping units "at levels of abstraction higher than the soil series." (They refer to this process as "categorical" generalization.) The effect of applying this second approach to taxonomic generalization is the same (whenever two similar mapping units share a common boundary, the boundary segment is eliminated to form a larger parcel), but it is achieved by a separate strategy. Olson et al (1980) applied this approach to produce a set of generalized statewide maps for Ohio, including single factor maps judged to be useful for understanding broad scale soil distributions.

Spatial Generalization. Spatial generalization refers to a third broad category of generalization strategies that combines those mapping units that tend to consistently occur adjacent to one another (Figure 8.4). Whereas graphic

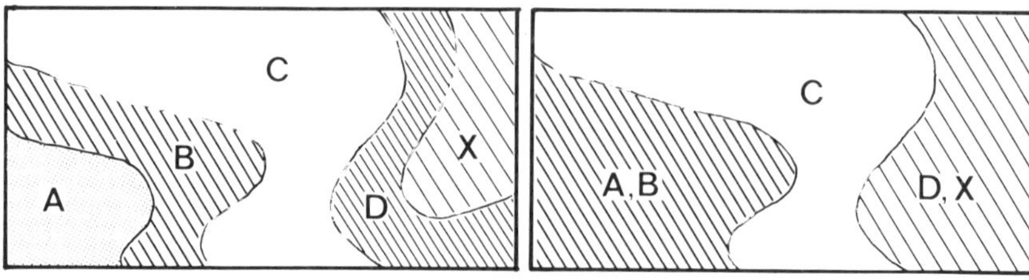

Figure 8.4 Spatial generalization combines mapping units known to consistently occur in adjacent landscape positions. The generalized map (*right*) simplifies the original detailed map (*left*) by combining diverse soils (eg: D,X) into composite mapping units. The reader can no longer be confident of the uniformity of the generalized mapping unit.

generalization proceeds by implementation of relatively easily defined rules that can be applied more or less independently of map content and the specific area mapped, spatial generalization requires a much more substantive knowledge of map content and specific geographic regions. In regard to soil maps, this form of generalization could also be referred to as "pedologic generalization," because it is less of a purely cartographic process than one requiring in-depth understanding of specific soil patterns.

Spatial generalization proceeds by deletion of boundaries between neighboring units, as does taxonomic generalization. However, spatial generalization joins greatly dissimilar soils, whereas taxonomic generalization aspires to preserve as much homogeneity within mapping units as may be possible. The critical difference is that specific pairs of taxa (for example), although pedologically distinct, may consistently occur in characteristic positions on the landscape, and with consistent neighbors. As a result, it is often convenient to group such soils together in a single composite mapping unit despite their differences.

The degree of simplification generated by taxonomic generalization depends upon the degree to which pedologically similar soils are contiguous to one another. It is conceivable that a pattern of contrasting soils that adjoin one another would experience only minor simplification if subjected to taxonomic generalization. Spatial generalization, on the other hand, allows grouping of dissimilar soils, so it can always produce a highly generalized map, at the expense of generating mapping units with high internal variation.

Spatial generalization is often based upon recognition of the major landscape units present within a region. These landscape units, and the consistent spatial arrangements of soils that may occur within them, are often related to repetitive geologic, geomorphic, hydrologic, or pedologic conditions. Although such units may exhibit recurring arrangements of soils, there is no reason to expect pedologic similarity. For example, Bleeker and Speight report that land systems in New Guinea exhibit only weak relationships between landforms and pedologic taxa, even at very broad levels: "... the most common great soil group of a land unit is rarely common enough to dominate the land unit ..." (1979, p. 189). Although situations vary greatly from those they encountered, their results highlight the pedologic diversity that can be present in mapping units defined on the basis of geographical propinquity.

As a result, spatial generalization requires a drastic alteration of the legend presented on the detailed map, in part because of the aggregation of pedologically distinct units and the necessity to specify as precisely as possible the composition of the generalized mapping units. Fridland (1976a) defines the following categories of generalized mapping units, ordered here according to the degree of specificity present in the legend. (Fridland's terminology, as presented in translation, is used.)

The *predominant soil* strategy relabels a composite region formed from several detailed mapping units with the identity of the areally predominant soil within the region. The presence of other soils can be symbolized, although normally such symbols would reveal only the presence of such soils, not their location or extent.

Soil cover composition generalization defines more clearly the character of a composite mapping unit by specifying its membership. In the United States, the soil complex represents the most indefinite form of soil cover composition generalization, as the content of the complex is specified, but the reader has no knowledge of proportions or patterns of member soils. The soil association is composed of soils in a consistent proportional membership and pattern, but the character of the pattern is not specified. There is no restriction on contrast within an association; membership may include soils belonging to taxa differing at the highest level in the classification.

Generalization by soil cover pattern specifies both proportions of member soils, and their distributions. For example, legends in a map generalized by this process might include "Brown forest soils on hills (30%); Podzolic soils on slopes (15%); Gleys on undrained plains (55%)."

Generalization by Sampling. Generalization by sampling is accomplished by labeling each unit in a network of areal units with a single category, determined by sampling the several categories actually present within each unit (Wehde, 1982). Assume that a systematic grid of uniform cells is superimposed over the detailed map; each cell is assigned a single soil designation based upon a category selected from those present within the cell (Chapter 6). This single category can be determined on the basis of areal dominance (i.e., selection of that category that occupies the greatest area within the

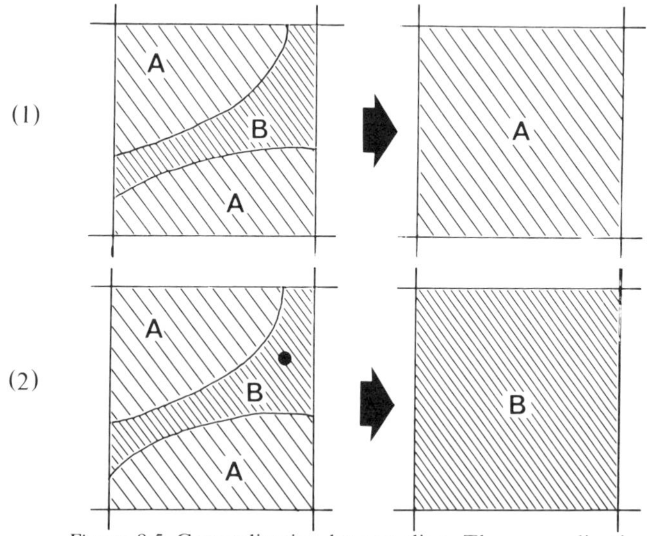

Figure 8.5 Generalization by sampling. The generalization of each cell (*right*) is assigned a single identity from those available in the detailed representation (*left*). 1: A single category is assigned to the cell by selecting the category that occupies the largest area within the cell. 2: The generalized category is assigned by selecting the category that falls beneath a dot randomly positioned within each cell.

cell), or by the position of a point located randomly within each cell (i.e., selection of the category that falls beneath the point, even though it may occupy only a small area). This general procedure is frequently used as a means of preparing soils information for entry into computer-based geographic information systems.

This kind of sampling procedure is itself a form of generalization that systematically favors certain kinds of soils for retention or exclusion on the basis of sizes, shapes, and locations of delineations (Nichols, 1975). Because generalization by sampling groups soils on the basis of location rather than upon pedologic similarity, it must be considered as a form of spatial generalization. Like all forms of spatial generalization, this procedure groups contrasting taxa together, generating imprecision and error. The amounts and distributions of error depend upon the size of the grid, and its relationship to the specific soil pattern to be mapped (Gersmehl and Napton, 1982).

In addition, generalization by sampling systematically excludes those soils that, due to characteristic sizes and shapes, cannot dominate a single cell (Figure 8.5–1). As a result, soils of pedologic significance that occur in long, narrow, delineations may be excluded from the sample. An alternative procedure is the use of a randomly positioned dot to select the category for each cell. However, although random selection improves the overall accuracy of the data base, it does not present information that permits assessment of the pedologic features of a region. For example, Figure 8.5–2 illustrates selection of a category of secondary areal significance–a result that may increase the accuracy as a whole, but one that prevents interpretation of the actual soil pattern from the sampled data.

Some of these problems might be avoided by subjecting the source map to spatial or taxonomic generalization prior to sampling. This provides the user with the identity of the generalized mapping unit (which will tend to occupy larger areas within cells), a more accurate indication of actual ground conditions than the single detailed category given by the usual procedures. Also, the coarser detail of the generalized map may be more nearly compatible with the broad scope of most geographic data bases.

FREQUENT USE OF SPATIAL GENERALIZATION

Examination of published soil maps reveals that many generalized maps seem to have been formed (in part at least) by spatial generalization. Individual large scale maps of most national soil surveys appear to have been subjected to graphic generalization (i.e., smoothing of boundaries, elimination of small delineations); examination of legends and accompanying documents often reveals the application of spatial generalization as well. County soil association maps of the U.S. Soil Conservation Service, and many of the state soil maps published by state agencies have been generated by grouping those detailed mapping units that occur together as landscape units. As a result, these generalized mapping units form useful representations of geomorphic regions and soil landscape units, but their pedologic diversity (some include members of as many as 3 soil orders) presents problems for those who attempt to use such maps as a source of pedologic information.

The usefulness of a soil map can be evaluated by the map user's ability to make precise statements about the composition of individual mapping units (Webster and Beckett, 1968). By this standard, mapping units formed by spatial generalization present problems for the map reader, since the diversity of mapping units can prevent use of the map as an accurate predictor of actual soil properties at specific points. These kinds of problems can, of course, be minimized by careful and complete description of mapping units.

INFLUENCE OF THE LANDSCAPE UPON GENERALIZATION

Specific landscapes will vary greatly in respect to complexity (and therefore in respect to their need for generalization), and also in their receptiveness to accurate generalization. We will review some of the features that most directly influence accuracy of generalization.

A landscape is composed of many patches of soil, each with distinctive pedologic characteristics, including the number and arrangement of horizons, texture, and depth, as well as a multitude of chemical, physical, and mineralogical properties. A fundamental premise of modern soil survey is that it is possible to identify and delineate, if only approximately, the outlines of discrete soil landscape units, and to correlate (recognize) individual soil species over fairly large distances. Each landscape unit can be characterized by hundreds, if not thousands, of pedologic properties, although only a few are normally reported in a soil survey. In addition to pedologic properties, each landscape unit also possesses a set of geographic properties, including size, shape, topographic position, slope, uniformity, etc (Hole, 1953). It is these geographic properties that are important in examining landscapes and the qualities that most strongly influence map generalization.

For example, individual delineations may be retained or excluded during generalization on the basis of size or shape, despite pedologic distinctiveness or significance for use and management. Lyford (1974) shows that long, narrow strips of soil in a New England forest, although of significance in forest management, were systematically excluded from soil maps, due to their sizes and shapes (Figure 3.15).

If the scope of examination expands from consideration of individual delineations to segments of the landscape composed of many delineations, a set of variables can be identified as significant for the generalization process. For a given area, the number (density) of individual delineations is of obvious significance, as are the number of distinct taxa represented, the sizes of delineations, and the pedologic contrast between them. If individual soils tend to form consistent associations, then presumably it will be easy to form composite mapping units based upon the soil association principle. If, on the other hand, spatial associations are absent or are weakly expressed, such units will be less effective. Habermann and Hole (1980) tabulate geographic properties for a variety of U.S. landscapes as a suggestion of the extent to which these geographic properties are likely to vary. Hole (1978) also presented examples.

GENERALIZATION ACCURACY AND GENERALIZATION PRECISION

Accuracy can be defined as the ability of a generalization process to assign areas on the generalized map to the same taxonomic groups in which they belong on the source map. (This definition assesses the accuracy of the process of generalization, but does not, of course, evaluate the accuracy of either the detailed or the generalized maps because there is no effort to compare either to the actual distribution on the landscape.) Possibly the most direct way to assess the accuracy of a generalization is to compare the generalized map, point by point, or area by area, with the more detailed source map. If they are presented at the same scale, the two maps can be superimposed to identify differences. For soil maps, one might examine the distributions of several soil properties on both patterns and thereby determine magnitudes and distributions of differences. The original detailed map will, of course, always have greater accuracy, but alternative generalization processes may yield differing amounts of error or differing spatial patterns of error. A generalization process that produces a few small differences ("errors") might be judged to be more satisfactory than one that yields larger or more widely distributed errors. The sizes, shapes, and patterns of such errors would also be significant in evaluating generalizations, so an error analysis should examine not only the frequency distributions, but also their spatial patterns.

Precison refers to the ability of a generalization to retain the same level of taxonomic detail portrayed on the original map. If a generalization were to retain the same boundaries from the detailed map, but identify delineations at higher levels in the classification, the accuracy of the map would not be altered, but the reader would have lost the ability to make precise statements about the locations represented on the map (see Orvedal and Edwards, 1941).

The person who prepares a generalized map assumes the responsibility for stating in the legend or in the accompanying report the composition of the mapping units formed by generalization. Thus, a legend for a generalized map should specify the approximate composition of each generalized mapping unit, and if possible, specify the patterns in which members occur within each mapping unit. If generalization includes smoothing of lines, it also seems reasonable to specify the presence of errors introduced by line generalization, and to identify the soil units influenced most by these errors.

ERROR PATTERNS

Since each generalization operation produces its own distinctive kind of error, it is possible to describe, in idealized form, the patterns of error likely

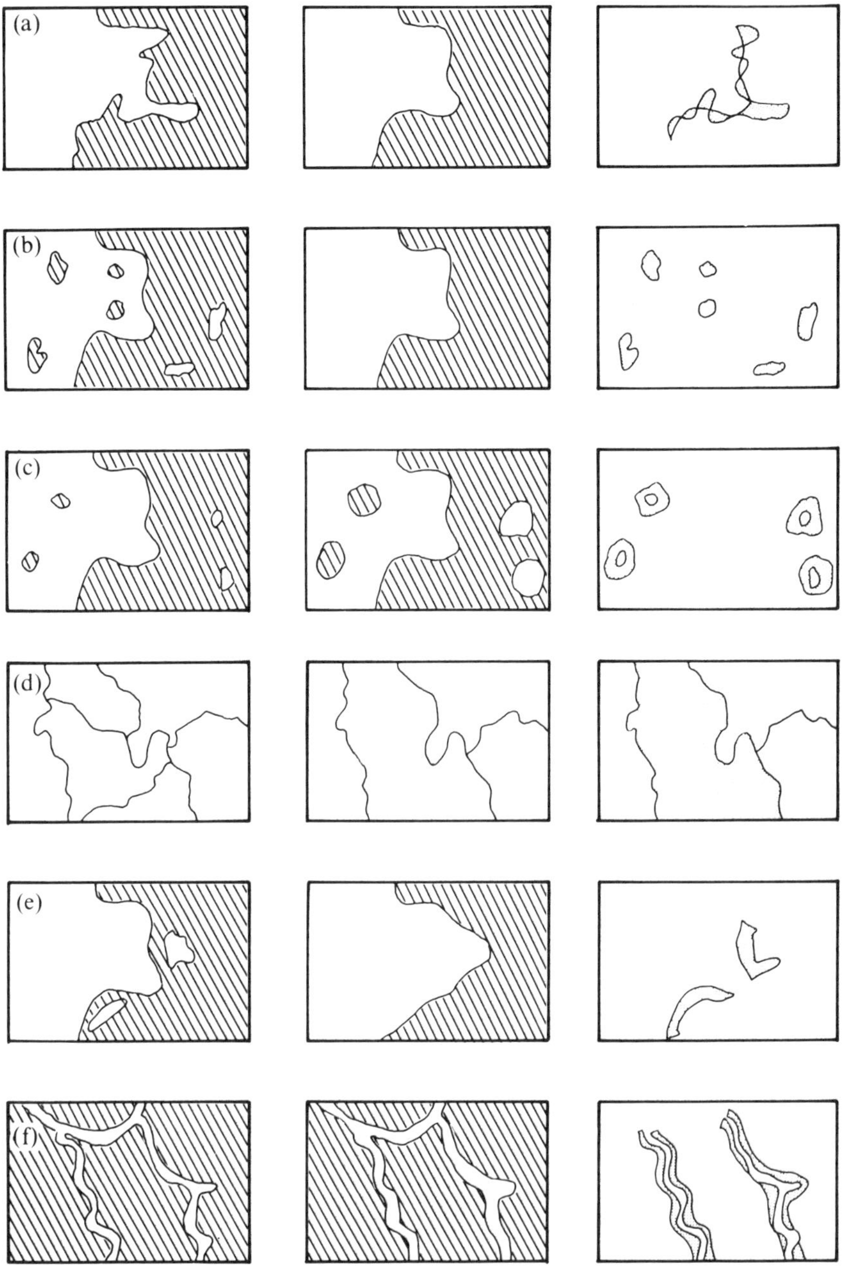

Figure 8.6. Idealized examples of generalizations of soil patterns. Each row shows a small section of a hypothetical soil pattern in its original detailed form (*left*), after generalization (*center*), and with generalization errors identified (*right*): (a) line smoothing, (b) deletion of boundaries, (c) enlargement of delineations, (d) boundary deletion, (e) addition of boundaries, and (f) shift in boundary position. Because most generalized maps have been simplified by two or more generalization processes, observed error patterns are combinations of several of these error distributions.

to be caused by a given generalization operation. Generalizations are usually made using a combination of the generalization operations described previously, so the actual errors on any given map will be formed by the accumulation of several of the patterns described here.

Boundary smoothing (Figure 8.6a) creates error on either side of the generalized line. Line smoothing algorithms transfer areas from a mapping unit to its neighbor, producing on the generalized map a chain of errors along the smoothed line. *Removal of delineations* (Figure 8.6b) causes a pattern of isolated islands of error, caused by the reassignment of the removed delineations to another mapping unit. *Enlargement of delineations* (Figure 8.6c) creates a zone of error bordering each of the original delineations. Errors form doughnut-shaped areas centered on each of the original delineations. *Deletion of boundary segments* (combinations of mapping units) (Figure 8.6d) generates on the generalized map zones of decreased precision throughout all areas occupied by both delineations. Whereas previously mentioned errors are local in extent, this operation generates a form of error that is more extensive, but less severe, and is documented in the legend. *Addition of boundary segments* (Figure 8.6e) creates a corridor of error between the two delineations, due to the transfer of strips (the shaded segments) from one category to another. Finally, *shifts in boundary positions* (Figure 8.6f) create bands of error that parallel the altered boundaries.

Many of the errors created by graphic generalization assume some practical significance if the map is to be sampled (for example) to select data for entry into a geographic data base. Some mapping units may be systematically altered in size or shape on the generalized map due to their characteristic sizes and shapes on the detailed map. Subsequent sampling of the generalization may incorporate these errors as omissions of certain soils, or as exaggerations of their areal significance in relation to their true areas on the ground.

Errors produced by spatial generalization vary greatly depending upon the form of generalization applied, the character of the landscape, and the organization of the legend. All spatial generalizations introduce imprecision, and may create error. Of the methods of spatial generalization, generalization by predominant soil produces the highest degree of uncertainty for the user, who is not informed of the true membership, proportions, or patterns within each mapping unit. Soil cover composition and soil pattern generalization present the reader with information concerning the true membership of mapping units, the relative proportions of soils within units, and (in the instance of soil pattern generalization) the arrangement of soils within mapping units. In general, spatial generalization requires a user to have at hand substantive knowledge of pedology and the geography of the mapped region.

PRINCIPLES OF GENERALIZATION

Several principles of generalization can be proposed a guidelines in preparing and interpreting generalized landscape maps.

1. The quality of a generalized soil map is a function of the specific landscape mapped and the generalization processes applied to the detailed map.

2. In principle, at least, a generalized map can be considered to be as rigorous as a detailed map in the sense that it should have known properties, including estimates of the amount of error present, and the assignment of error to various mapping units, and locations on the generalized map.

3. Although there is no point in attempting a complete accounting of generalization error (this would in effect reconstruct the detailed map), sound cartographic practice requires some effort to inform the user of specific merits and deficiences of each generalized map, and of mapping units within each map. In the United States, current county soil association maps, for example, provide descriptions of the (nominal) composition of generalized mapping units, yet probably could be improved by more explicit recognition of the contrast present within such units, and of the fact that composition is not uniform throughout each delineation.

4. Some pedologic situations inherently defy convenient generalization into homogeneous mapping units. For example, soils on stream terraces may display a strong contrast with adjacent alluvial soils. The complex spatial arrangement prevents formation of homogeneous generalized mapping units, so attempts to generalize must inevitably create extremely heterogeneous mapping units.

5. The overall generalization problem is essentially one of partitioning diversity into broad categories, subject to the spatial constraints imposed by the sizes, shapes, and arrangements of detailed mapping units.

Epilogue

Final Comment on Soil Landscape Dynamics

Five things influence soil landscapes;
Biota, climate, terrain shapes,
Initial stuff and human kind.
A blend of these may loose or bind
The land skin of hills and dales:
Here turns soil dark, and elsewhere pales;
Leaches it poor or makes it rich,
Defining each natural niche,
Erodes or catches soil debris,
Changing the landscape endlessly.

Appendix I

PROVISIONAL SOIL SLOPE AND SOIL CONTRAST MAPS OF THE U.S.A.

The SOIL SLOPE REGIONS map (Figure I-1) is so designated because the information was derived from the soil map of the U.S. (Sheet No. 86 in the *National Atlas*) (Soil Survey Staff, 1967), and the slopes are reported in percent gradients. Slope is a property of soil bodies. The natural atlas soil map is, of course, itself a rather broad generalization, so this map should not form the basis for detailed, site-specific interpretations, as it is presented at a scale much smaller than that of the source map.

The provisional map of CONTRAST OF THE SOIL COVER PATTERN (Figure I-2) is exploratory in nature. It was compiled by noting the number of soil orders listed for each delineation in the *National Atlas* soil map. Interpretative license was necessary in making decisions for classification of borderline cases. Generalization and interpretation was necessarily subjective, so the result can be considered only as a first approximation. The three major categories of soil contrast within delineations may be illustrated by the three following examples (1U, 2U, and 3U, respectively), with corresponding symbols from the *National Atlas* map indicated:

E10-2—Quartzipsamments plus Paleudults, gently or moderately sloping.
S4-4—Haplorthods plus Fragiorthods, moderately sloping.
A7-1—Hapludalfs, gently sloping.

No wetland soils are listed for these three units. Unit S1-1—Haplaquods plus Quartzipsamments, gently sloping, contains both poorly drained and excessively drained soils, and so is designated 1W in the figure. No body large enough to delineate was found for category 2W in the legend; it is quite possible that future work will discover such a body or bodies. Bodies occupied almost entirely by wetland soils are illustrated for 1WW, 2WW and 3WW categories by the following units from the *National Atlas* soil map:

A2-3—Ochraqualfs plus Psammaquents, gently sloping.
U1-4—Ochraquults plus Umbraquults, and tidal marsh, gently sloping.
H2-1—Histosols (plant residues moderately decomposed), gently sloping.

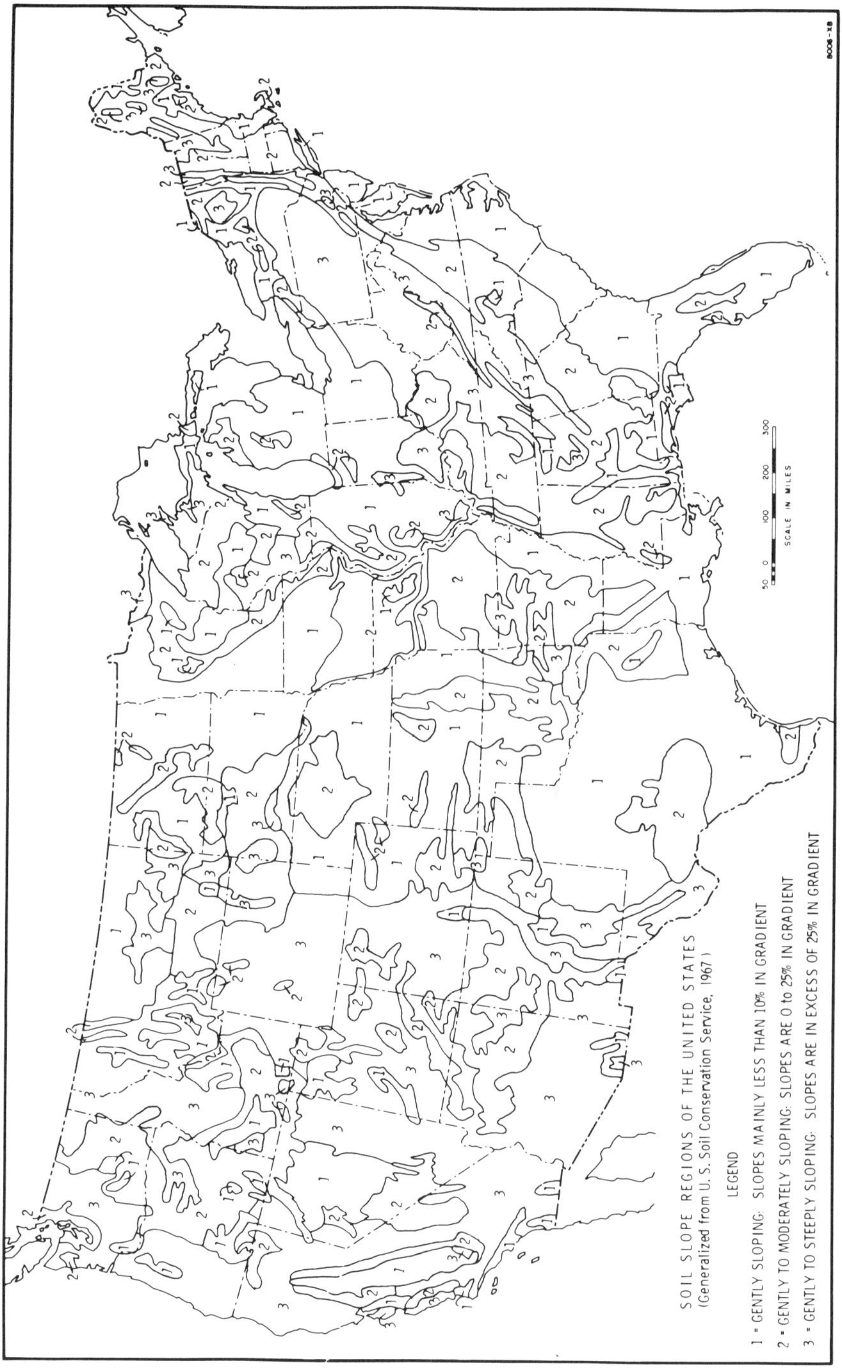

Figure I-1. Provisional soil slope regions of the United States.

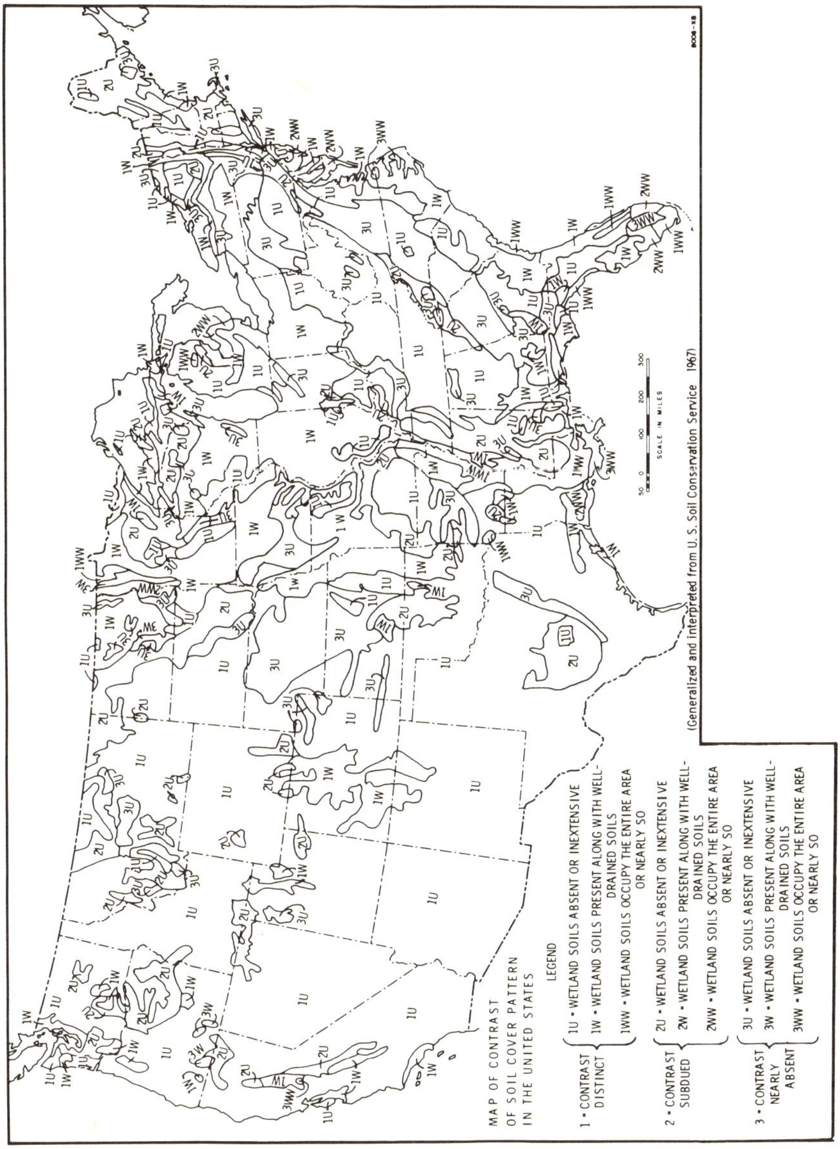

Figure I-2. Provisional soil contrast regions of the United States.

Appendix II

PROCEDURE OF SOIL MAP ANALYSIS FOR CHARACTERIZING SOIL LANDSCAPES IN TERMS OF HYDROLOGY OF SOIL COVER

The application of the *Soil Taxonomy* (Soil Survey Staff, 1975a) in routine soil mapping in the United States has made available in greater detail than ever before information about soil sequences along the stream-line on floors of valleys, as well as on the enclosing valley walls. Therefore, it is now possible to study the linkage between soil sequences along these two different kinds of transects, using the wealth of information in modern published soil surveys, with additional observations from pertinent ground checks.

The procedure of soil map analysis described here is illustrated by Figures II-1 through II-3 and Tables II-1 through II-3 with respect to characterization of stream-line and valley-wall catenas, and by Figures II-4 and II-5 for for pedohydrologic characterization of soil survey map sheets covering about 8 sq. km each.

Characterization of stream-line and valley-wall catenas. A schematic block diagram of an upper segment of a valley is shown in Figure II-1A. Along the drainage line are eight soil bodies. The up-stream boundaries of bodies numbers 2 through 8 are called pedological nodes, because they are used as points of origin of transects up the valley wall onto the adjacent upland. The soils encountered along these transects constitute the sequences called catenas (topo-litho-hydro sequences), of which seven are shown in the figure. The up-stream boundary of soil body number 8, a very poorly drained organic soil, is counted as the last node. This is an arbitrary selection of a criterion for limiting the study to the upper reaches of the drainage-way where relationships between soils of the upland and of the valley are most intimate.

Figure II-1B presents a statistical representation of information derived from the soil map of the valley segment shown in Figure II-1A. Natural soil drainage indices, on a scale of 1 to 100, are arbitrarily assigned to the natural drainage conditions of the soils, which are interpreted from the taxonomic classification of the soils, This is illustrated by the following representative list:

Taxonomic classification of some representative soils	*Clues to the moisture regime present*	*Natural soil drainage term*	*Natural soil drainage index*
Smooth granite outcrop. No soil classification	No soil at all. Nearly impermeable bedrock	None	0
Lithic Udipsamment, sandy (unmottled)	A sandy (psamm) soil shallow over bedrock (Lithic) is xerix even in a humid (ud) area	Very excessively drained	10
Lithic Hapludalf, fine-silty (unmottled)	Silt loam shallow to bedrock in a humid region	Excessively drained	20
Typic Hapludalf, fine-silty (unmottled in first meter)	Deep silt loam in humid region, with no mottling to indicate impedency of drainage	Well drained	40
Typic Hapludalf, fine-silty (slightly mottled below 0.8 meter)	Same as preceding soil except mottling indicates some seasonal impedence of drainage	Moderately well drained	50
Aquic Hapludalf, fine-silty (mottled as high as A2 horizon, at 10 cm; bluish-gray "gleyed" colors notable only at depth of 1 meter)	Mottling in entire solum indicates seasonal high water table	Somewhat poorly drained	60
Aeric Fragiaqualf, fine-silty (mottled as high as 10 cm below surface; considerable bluish-gray "gleyed" color below 50 cm depth)	Notable bluish-gray "gleyed" colors indicate longer duration of high water table than in the preceding soil	Somewhat poorly to poorly drained	70
Typic Haplaquoll, fine-silty (bluish-gray "gleyed" B horizon under a thick black A1 horizon)	Accumulation of organic matter in silty surface soil, and gleying in B indicate long-term seasonal high water table	Poorly drained	80

Histic Haplaquoll (a 250 cm thick peat layer overlies the black A1 horizon and gleyed B horizon)	Accumulation of peat indicates that this soil has poorer drainage than the preceding soil	Poorly to very poorly drained	90
Typic Borofibrist (a saturated peat in a northern climatic zone, such as that adjacent to Lake Superior)	Two meters of wet peat indicate that a high water table exists through most of the year	Very poorly drained	100

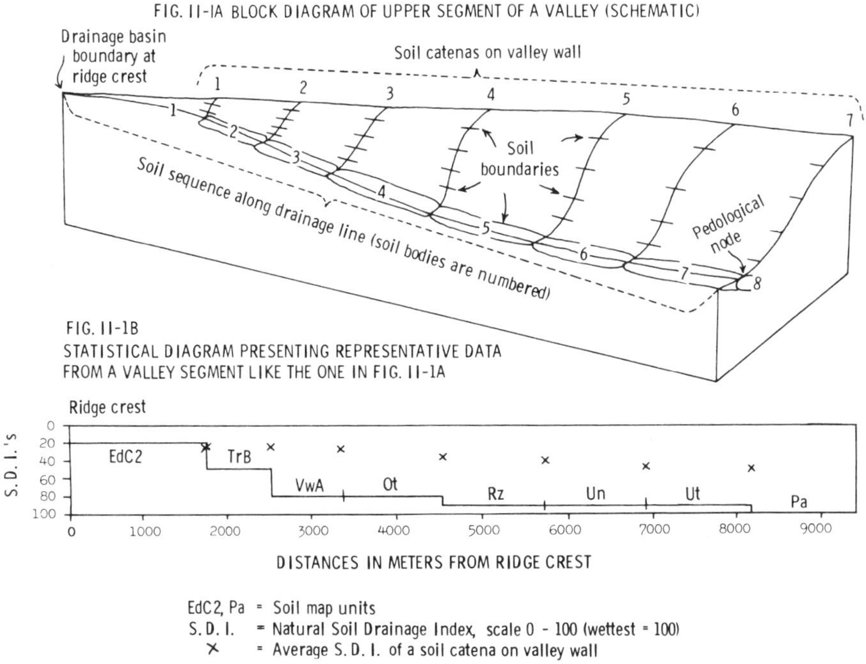

Figure II-1. Schematic block diagram (1A) illustrating the numbering of drainage-line soil bodies and related catenas in a valley segment, and a statistical diagram (1B) of representative data in terms of natural soil drainage index.
EdC2, Pa = Soil Map Units
S.D.I. = Natural Soil Drainage Index (scale 0-100, wettest = 100)
X = Average S.D.I. of a soil catena on valley wall.

Figure II-1B shows that the drainage-way crosses, in succession, an excessively drained (droughty) soil (EdC2), a moderately well drained soil (TrB), two poorly drained soils (VwA, Ot), three poorly to very poorly drained soils (Rz, Yn, Ut) and a very poorly drained soil (Pa). The trend down-valley is from bodies of usually unsaturated soils to bodies of saturated soils. The figure also indicates that the average drainage condition of soils of catenas on adjacent valley walls in the lower two-thirds of the valley segment is about 50 units drier than the nearest valley-bottom soils. The upland is clearly well drained; the valley bottom becomes progressively less well drained.

In Figure II-2, statistical data are plotted for another valley segment that is in many repects similar to the schematic diagram of Figure II-1A. The interruption of the trend of increasing wetness down-valley by bodies of Os soil is a result of thickened deposits of alluvium by tributaries. Data from a valley segment in Ashtabula County, Ohio are plotted in Figure II-3. These reveal a landscape in which the upland is largely occupied by poorly drained soils. The trend down-valley is for the valley floor to become better drained, attaining a moderately well drained condition at soil body Lb.

More information for the two valley segments just discussed and illustrated is presented in Table II-1. The average natural soils drainage index (S.D.I.) for the catenas are 38.5 and 72.4 for the Wisconsin and Ohio segments, respectively, whereas the averages for the drainage-way soils are almost identical. The ratios between the S.D.I. figures in each instance are revealing as to the linkage between soil moisture regimes in the two principal kinds of traverses. A value less than one indicates an upland better drained than soils of the drainage-way. A value more than one indicates a reverse condition.

FIG. II-2

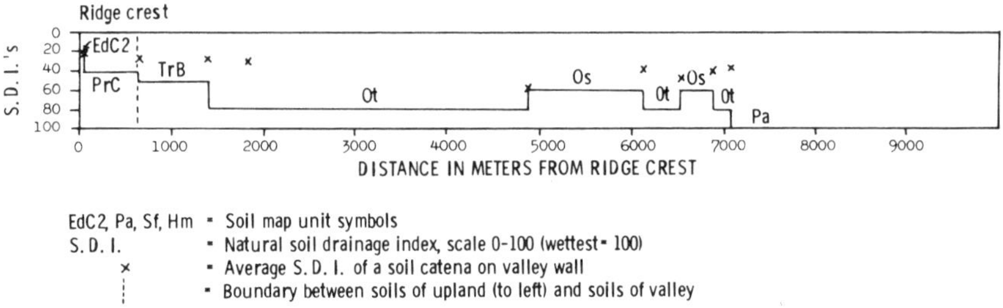

Figure II-2. Statistical diagram presenting data from a valley segment in Dane County, Wisconsin, U.S. (Glocker and Patzer, 1978).
EdC2, Pa = Soil Map Units Symbols (Glocker and Patzer, 1978)
S.D.I. = Natural Soil Drainage Index
X = Average S.D.I. of a soil catena on valley wall. | = Boundary between soils of upland (left) and valley (right)

Table II.1 Information on the Nature of Soil Sequences Along Drainage Ways and on Adjacent Valley Walls in Selected Landscapes of Two Representative Counties of the North Central Region of the United States.

	Soil sequences down stream-line					Soil sequences down slopes of valley walls						Summary Values	
County and State	Total Length (m)	Total Relief (m)	Av. Gradient[a] (%)	Av. soil drainage index[b] (SDI)	Ratio, SDI of first to last (lowest) soil bodies	Av. Length (m)	Av. Relief (m)	Av. Gradient (%)	Av. No. parent materials per catena	Av. No. soil bodies per catena	Av. SDI of catena	Av. SDI for all soils	Ratio, Av. SDI for catenas to that of drainage way soil
Dane, WI	7,178	75	1.04	67.4	0.2	317	25	7.8	2.3	4.3	38.5	59.2	0.57
Ashtabula, OH	10,422	30	0.3	67.9	1.2	305	5	1.5	1.4	4.6	72.4	68.9	1.07

Notes: [a] Gradient was calculated on the basis of Figures II.2 and II.3.

[b] Natural soil drainage index (SDI) is obtained by assigning arbitrary scaled values to natural soil drainage conditions: Excessively drained, 10 (as on a sand dune); Well drained, 40; Moderately well drained, 50; Somewhat poorly drained, 60 (as in Aquic Hapludalf); Somewhat poorly to poorly drained, 70 (as in Aeric Fragiaqualf); Poorly drained, 80; Very poorly drained, 100 (as in peat or muck: Histosols). Average SDI for a soil sequence is determined as the product of length of transect segment across a soil body and the SDI for that body, with summation for all soil bodies in the sequence, divided by the total length of the sequence.

Table II.2 Soil Landscape Information for Uppermost Reaches of a Tributary of the Sugar River, Dane County, WI

Soil sequences down stream-lines[a]				Soil sequences down slopes of valley walls[b]					
Soil map symbol[c]	Soil wetness index[d]	Horizontal distance between soil boundaries (m)	Vertical distance between soil boundaries (m)	Soil map symbol[c]	Soil wetness index[d]	Horizontal distance between soil boundaries (m)		Vertical distance between soil boundaries (m)	
EdC2	20	31.4	3.0	NeB2	35	30	202	1.5	4.6
				EdC2	20	172		3.1	
PrC	40	627.2	24.4	NeB2	35	30	233	3.1	27.5
				EdC2	20	157		18.3	
				PrC	40	46		6.1	
TrB	50	752.0	21.3	NeB2	35	30	308	3.1	36.6
				EdC2	20	141		15.2	
				NeD2	30	93		9.1	
				PrC	40	30		6.1	
				TrB	50	14		3.1	
VwA	80	424.0	6.1	DuB2	35	30	372	3.1	42.8
				DuC2	30	141		18.3	
				NeD2	30	157		15.2	
				HbC2	30	30		3.1	
				VwA	80	14		3.1	
Ot	80	3008	16.8	HbC2	30	78	337	3.1	15.4
				HbD2	25	30		1.5	
				GaC2	35	30		3.1	
				SmB	40	30		3.1	
				VwA	80	30		1.5	
				Os	60	30		0.0	
				Ot	80	109		3.1	
Os	60	1282.0	1.5	GaB	35	109	389	4.6	16.9
				EhC2	25	109		4.6	
				GaB	35	78		4.6	
				Os	60	93		3.1	

Ot	80	406.0	1.5	EdC2	20	93	265	10.7	25.9
				RaA	55	78		10.7	
				Os	60	78		4.5	
				Ot	89	16		0.0	
Os	60	360.0	NM	EdC2	20	46	326	4.6	26.0
				EdD2	15	98		9.1	
				RaA	55	78		7.7	
				Os	60	104		4.6	
Ot	80	187	NM	EdC2	20	62	422	4.6	26.0
				EdD2	15	173		12.2	
				RaA	55	125		6.1	
				Ot	80	62		3.1	
(Pa)	100								

[a] Modern soil maps using the new USDA Soil Taxonomy report in greater detail than ever before the soil sequence on valley floors.
[b] These sequences are referred to in the literature as the soil catenas.
[c] The soil map symbols correspond to the following list of soils. Slopes are A (or no slope symbol at all) = 0-2%; B = 2-6%; C = 6-12%; D = 12-20%. Degrees of erosion are: - = slight or none; 2 = moderate; 3 = severe.
Du — Dunbarton silt loam. Clayey montmorillonitic mesic. Lithic Hapludoll; Ed — Edmund silt loam. Clayey montmorillonitic mesic. Lithic Argiudoll. Eh — Eleva sandy loam. Coarse-loamy mixed mesic. Typic Hapludalf; Ga — Gale silt loam. Fine-silty over sandy or sandy-skeletal mixed mesic. Typic Hapludalf. Hb — Hixton loam. Fine-loamy over sandy or sandy-skeletal mixed mesic. Typic Hapludalf; Ne — New Glarus silt loam. Fine-silty over clayey mixed mesic. Typic Hapludalf; Os — Orion silt loam. Coarse-silty mixed nonacid mesic. Aquic Udifluvent; Ot — Otter silt loam. Fine-silty mixed mesic. Cumulic Haplaquoll; Pa — Palms muck. Loamy, mixed, euic, mesic. Terric Medisparist. (Note the stream passes by this body of soil, not through it); Pr — Port Byron silt loam. Fine-silty mixed mesic. Typic Hapludoll; Ra — Radford silt loam. Fine-silty mixed mesic. Fluventic Hapludoll; Sm — Seaton silt loam. Fine-silty mixed mesic. Typic Hapludalf; Tr — Troxel silt loam. Fine-silty mixed mesic. Typic (Fluventic) Argiudoll; Vw — Virgil silt loam. Gravelly substratum. Fine-silty mixed mesic. Udollic Ochraqualf.
[d] Arbitrary scaled values are assigned to the various natural soil drainage (dryness-wetness) conditions, as follows (Hole, 1978): Excessively drained, 10; Well drained, 40; Moderately well drained, 50; Somewhat poorly drained, 60; Poorly drained, 80; Very poorly drained, 100.

Source: Center section 10, T 6 N, R 7 E, 7177.6 Photosheet 74.6 No. 111 in Glocker and Patzer (1978). The 2854 stream line begins at the 221.7 northern edge of sheet No. 123 in soil body EdC2, 1,000 feet northeast of highway 151-18. Draining northward, it crosses onto sheet No. 111, turning northeastward, crossing onto sheet No. 112. It begins to flow east and then southeast near a road junction. Water follows a man-made ditch straight east to a channelized portion of the Sugar River, and continues about 2,000 feet before it parallels a body of Palms muck. See Glocker, Carl L., and Robert A. Patzer (1978). Soil Survey of Dane County, Wisconsin, U.S.D.A. Soil Conservation Service. Superintendent of Documents, Washington, D.C., 20402. 193 pages. 181 photo-soil-map sheets.

Table II.3 Soil Landscape Information for Uppermost Reaches of a Tributary of the Grand River, Ashtabula County, OH.

Soil sequences down stream-lines[a]				Soil sequences down slopes of valley walls[b]					
Soil map symbol[c]	Soil wetness index[d]	Horizontal distance between soil boundaries (m)	Vertical distance between soil boundaries (m)	Soil map symbol[c]	Soil wetness index[d]	Horizontal distance between soil boundaries (m)		Vertical distance between soil boundaries (m)	
Psa	60	63	NM[e]	Psa	60	63	63.0	NM	NM
Sf	80	189.2	NM	Psa	60	47.3	204.9	NM	NM
				Sf	80	157.6		NM	
Psa	60	78.8	NM	Psa	60	47.3	297.5	NM	NM
				Sf	80	157.6		NM	
				Psa	60	94.6		NM	
Sf	80	2445.0	9.1	Psa	60	31.5	473.1	NM	3.0
				Sf	80	441.6		3.0	
Hm	80	563.3	3.0	Psa	60	47.3	283.8	NM	3.0
				Sf	80	110.4		NM	
				Psa	60	110.4		1.5	
				Hm	80	15.7		1.5	
Os	60	820.3	3.0	Sf	80	236.7	362.8	4.6	6.1
				Os	60	126.1		1.5	
Hm	80	962.2	6.1	PsB	60	31.5	378.5	NM	9.1
				Sf	80	268.2		4.6	
				PsC2	55	63.0		3.0	
				Hm	80	15.8		1.5	

WIB	40	47.3	NM	PsB	60	31.5	410.0	NM	9.1
				Sf	80	283.9		3.0	
				PsC2	55	47.3		NM	
				WIB	40	47.3		6.1	
Os	60	5252.9	9.1	Sf	80	78.8	267.9	1.5	12.1
				PsB	60	141.9		9.1	
				Lb	50	31.5		1.5	
				Os	60	15.7		NM	
Lb	50								
		10422	30.3				2741.5		42.4

[a] Modern soil maps using ***Soil Taxonomy*** report in greater detail than ever before the soil sequence on valley floors.

[b] These sequences are referred to in the literature as soil catenas.

[c] The soil map symbols correspond to the following list of soils (Slopes are A or no slope symbol at all = 0–2%; B = 2–6%; C = 6–12%; d = 12–18%; Degrees of erosion are — = slight or none; 2 = moderate; 3 = severe):
Hm — Holly silt loam. Fine-loamy mixed nonacid mesic. Typic Fluvaquent; Lb — Lobdell silt loam. Fine-loamy mixed mesic. Fluventic Eutrochrept; Os — Orrville silt loam. Fine-loamy mixed nonacid mesic. Aeric Fluvaquent; Ps — Platea silt loam. Fine-silty mixed mesic. Aeric Fragiaqualf; Sf — Sheffield silt loam. Fine-silty mixed mesic. Typic Fragiaqualf; WI — Williamson silt loam. Coarse-silty mixed mesic. Typic Fragiochrept.

[d] Arbitrary scaled values are assigned to the various natural soil drainage (dryness-wetness) conditions, as follows (Hole, 1978): Excessively drained, 10; Well drained, 40; Moderately well drained, 50; Somewhat poorly drained, 60; Poorly drained, 80; Very poorly drained, 100.

[e] NM = Not Measurable on the U.S. Geological Survey topographic quadrangles. The soil map indicates that the slope is less than 2% (2 feet of fall per hundred feet of horizontal distance).

Source: The drainageway begins in a body of soil Psa in the extreme north central part of photo-map sheet No. 40, trends northward onto sheet No. 33, where it comes to parallel the Penn Central R.R. tracks, crossing onto sheet No. 26, whence it meanders back onto sheet No. 33, then to sheet No. 32, onto sheet No. 25, and back onto No. 32 to sheet No. 25 again where it enters Mill Creek.

Tables II-2 and II-3 in the Appendix tabulate the data for the two valley segments in question, and list the taxonomic classification of each map unit, thereby explaining the soil symbols (EdC2, Pa, etc.). The graphs of Figure II-2 and 3 and the indices of Table II-1 by-pass the taxonomic nomenclature and characterize the soil landscapes clearly, preparing the way for correlation of soil conditions with other phenomena, such as hydrogeologic conditions (Bouma and Hole, 1971) and effects of activities of human beings.

Characterization of pedohydrology of samples of soil landscapes. By means of transects across one or more soil photo-maps, the mean soil drainage index and histograms of soil drainage conditions in the two landscapes just considered (Table II-1) may be determined. Data were collected at 1 cm intervals along diagonal transects connecting the corners of the soil photo-maps identified in the caption for Figures II-4 and II-5. The characterization of the entire landscape takes into account the geographic extent of soils of

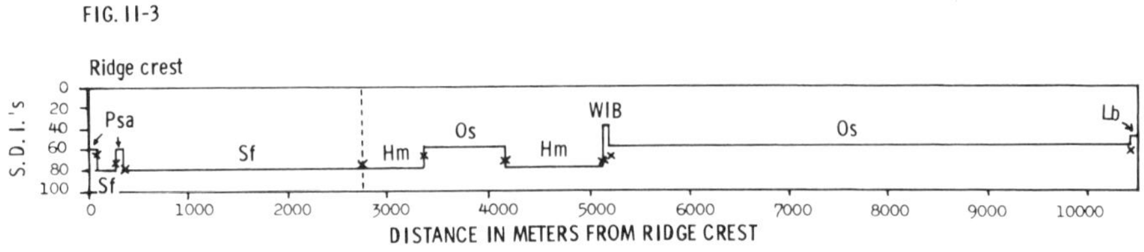

Figure II-3. Statistical diagram presenting data for a valley segment in Ashtabula County, Ohio, U.S. (Rieder, Riemenschneider and Reese, 1973).
EdC2, Pa = Soil Map Units Symbols (Reeder et al 1973)
S.D.I. = Natural Soil Drainage Index
X = Average S.D.I. of a soil catena on valley wall. | = Boundary between soils of upland (left) and valley (right)

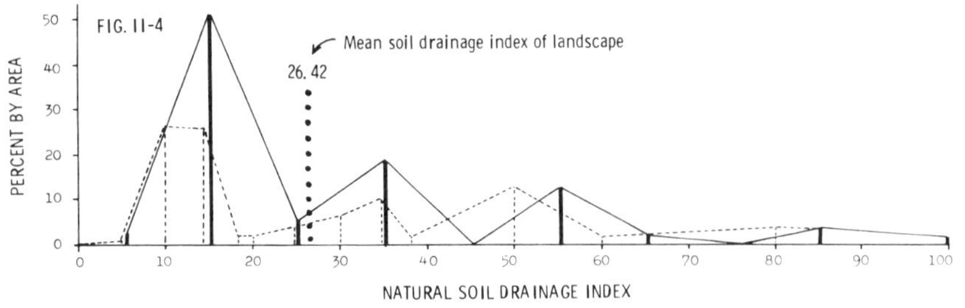

Figure II-4. Plot of analysis of soil photo map number 111, Dane County, Wisconsin, U.S. (Glocker and Patzer, 1978). Dashed lines represent original data, solid lines are generalized summary curves.
Note: Solid bold histograms are summarized by decades. Dashed histograms report soil phase by soil phase. Bold dotted vertical line is the mean soil drainage index.

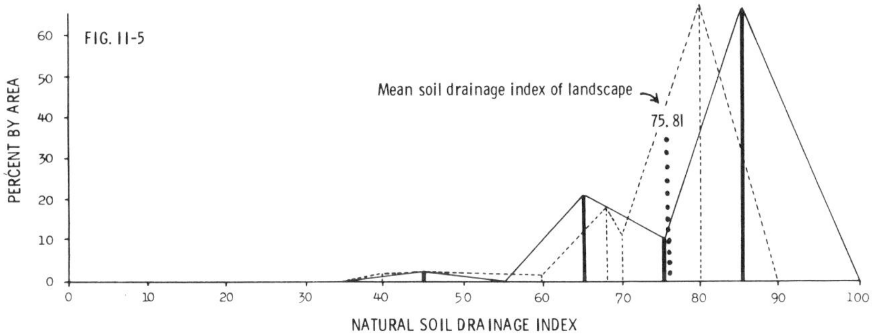

Figure II-5. Plot of analysis of soil landscape of soil photo maps numbers 33 and 34, Ashtabula County, Ohio, U.S. (Rieder, Riemenscheider, and Reese, 1973). Dashed lines represent original data. Solid lines are generalized summary curves. Note: Solid bold histograms are summarized by decades. Dashed historgrams report soil phase by soil phase. Bold dotted vertical line is the mean soil drainage indes.

different drainage categories on the interfluves, as the data of Table II-1 and Figures II-2 and II-3 did not. The mean soil drainage index for the Wisconsin landscape (Figure II-4) is 26.42 as compared with 38.5 for soil catenas on valley walls of one tributary (Table II-1), indicating that excessively drained soils dominate in the upland areas. The mean soil drainage index for the Ohio landscape (Figure II-5) is 75.81, which is close to the 72.4 index for catenas on valley walls (Table II-1), indicating that poorly to somewhat poorly drained soils dominate the interfluves.

The two landscapes in question, the one in Wisconsin and the one in Ohio, may be generally and tentatively defined as (1) one (Wisconsin) in which water moves from upland surfaces rapidly by a combination of infiltration and surface runoff, recharging the groundwater, and bringing the water table to the surface, as revealed by the soil map, in bodies of peat and muck about 8,250 meters distant down-stream from the ridge-crest; (2) and one (Ohio) in which water is stored in the upland soil bodies as well as in the underlying geologic materials (below a depth of 2 m), and into which the soil water moves slowly, continually by unsaturated flow and periodically by limited saturated flow. Downward movement of water is impeded by a fragipan in the silty profiles of the Sheffield and Platea soils (Ohio). Probably most rapid movement is about 3.0 cm per day at saturation and 1 cm per day at 10 MB AR tension (Figure II-6), with corresponding travel-times to move water 30 cm of 10 days and 30 days, respectively (Figure II-7). By contrast, the Dunbarton and Edmund soils of the Wisconsin landscape have conductivities ten times those just mentioned and travel times one-tenth as long. The Wisconsin landscape has much more surface runoff than the Ohio landscape because of greater relief in the former, in which relatively rapid infiltration results from heterogeneity of the soils. This refers in part to discontinuity of the residual clay layer in the subsoil, which means that "windows" are present in the soil cover through which water may move more rapidly than might by indicated by reference to the modal soil profile descriptions alone.

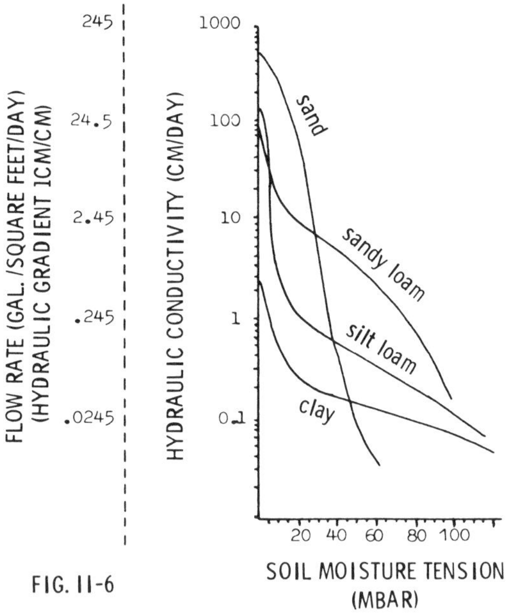

FIG. II-6

Figure II-6. Hydraulic conductivity (K) for some major soil horizons, of the textures indicated, as a function of soil moisture tension measured *in situ* with the crust-test procedure (after Figure 1 in Bouma, 1974).

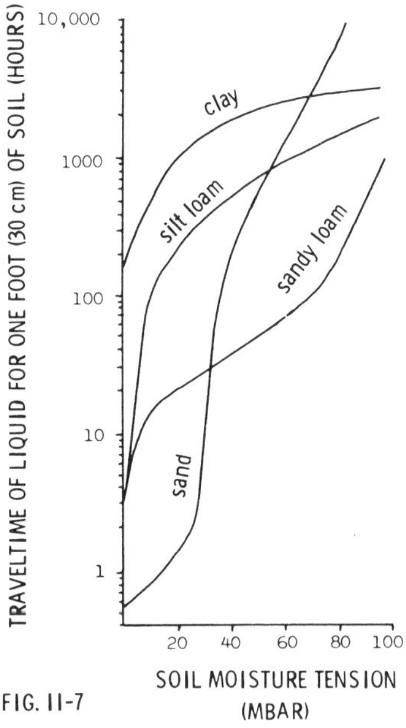

FIG. II-7

Figure II-7. Traveltime of liquid for one foot of soil at different soil moisture tensions, calculated for four soil materials. (Figure 6.7, Bouma et al, 1972).

Glossary of Terms

Alluvial toeslope position. A concave upward slope at the lowest part of a hillside, below the footslope position (Rhue, 1969).

Analogous soil combinations. These combinations have the same content with respect to soil taxa and their proportionate extents, but have been produced by different genetic steps, and therefore have different interrelationships. Example: Similar complexes in the Caspian lowland and Ukraine developed under subsidence of the water table in the first instance, and the reverse in the second instance. The complex consists of solonetzes on rises with meadow-chestnut soils in depressions. See *Homologous series of soil combinations.*

Anisotropic. A condition of a soil profile of being layered (horizonated) and in that sense, regularly heterogenous, in contrast to the isotropic nature of the original initial material. In the context of place-to-place variability, "anisotropic" refers to the fact that some soil characteristics may vary at different rates as direction varies. That is, place-to-place changes in the soil cover (within a specific landscape unit) may not be uniform, as different directions are considered. (See Campbell, 1978; Zaslavsky and Rogowski, 1969; Walker, Hall, and Protz, 1970).

Anthropic changes in soil cover. Changes made by human activity, including the following changes in the soil cover: (1) diversification as a result of erosion and related sedimentation; (2) homogenization resulting from agricultural practices that raise levels of crop yields. Included are changes caused by activities such as land-forming, land-leveling, deforestation, and creation of reservoirs (Fridland, 1976a, p. 81).

Apedal. Condition of a soil that has no structure, i.e., no peds, but rather is massive or composed of single grains. See *Pedal.*

Areal. A geographical extent in the sense of limit of range of a species. The term has been used in this way by German botanists, as illustrated by the title: "Arealkunde, floristisch-historisch Geobotanik," by H. Walter and H. Straka (1970; Stutgart, 478 pp.). An areal is the surface of land within the limits of which one finds one or another species of a soil, plant, or animal. "Elementary soil areal" is a different concept (see below).

Artificial soil body. A soil body that has been consciously, explicitly constructed by mankind's actions. Examples include the making of artifical sequences of horizons in land reclaimed from strip-mined areas, and similar activities to build manmade soils in lands reclaimed from the sea.

Generally excluded are natural soils that have been altered by mankind in a progressive or accidental manner.

Artificial soil individual. A human construct in a continuous universe (Knox, 1965). See *Natural soil individual.*

Autometamorphosis. Change in the soil cover as a result of internal cumulative alteration. For example, podzolization may proceed until the B horizon becomes a moisture barrier and pseudogley conditions result; homogenization may take place above the B horizon. Humidification promotes autometamorphosis. Contrast with *Parametamamorphosis* (Fridland, 1976a, p. 104).

Autonomous soil. This is soil that has formed exclusively under the influence of atmospheric moisture (Fridland, 1976, p. 85). See *Heteronomous soil.*

Background soil body, or punctate soil body. A soil body that encloses other elementary soil bodies within it. See *Punctate soil body* and *Punctate patchwork.*

Backslope position. The free face of a scarp of a hillside that lies below the shoulder and above the footslope (as seen in cross-section) (Ruhe, 1969; 1975). This corresponds to the transportational midslope of Conacher and Dalrymple (1977).

Bonding. Interaction between neighboring soil bodies. See *Dynamics of a soil combination.*

Bootan. Local Russian term for molehill, 20–50 cm high, in undrained portions of the Caspian plain (Fridland, 1976a, p. 141).

Bor. A type of sparse pine forest growing on sandy soil such as dune sand. This local Russian term may be extended to any pine forest (Fridland, 1976, p. 184).

Buried soil body. A soil body that has been buried under one or more deposits of sediments or artificial pavement.

Byndell. A group of parallel catenas (Bushnell, 1942).

Catena. A combinational soil body in which the "distribution of the soils is a function of level (elevation) and slope" (Milne, 1935). Two kinds of catenas are: (1) those on uniform geologic materials; (2) those on contrasting stratified rock layers. The French *chaine des sols* concept included the idea of a genetic interrelationship between soil members. Bushnell (1942) conceived of the catena as a topohydrosequence of soils developed in a single parent material, such as a glacial till. A body of peat on the till in a footslope position was not considered to be a different parent material. Fridland (1976a, p.11) has noted the varied use of the term.

Catenic Soil Cover Pattern. A system on a soil landscape universe (a soil combination) of interrelated repeating components (Fridland, 1976a, p.6).

Chaine des sols. See See *Catena.*

Chorosequence of Soils. A sequence of soil through space, as along a transect across terrain. Toposequence refers to a chorosequence across terrain with relief.

Chronosequence of Soils. An array of soil bodies on terrain, the principal differences among which may be ascribed to differences in age. Contrast with *chorosequence.*

Class. "An abstract field created by a class concept (which) expresses the basis for membership in it (in terms of) one or more differentiating characterisitics" (Knox, 1965).

Classificational ranks of soil cover. These are (1) elementary soil body, (2) simple combinational soil body, (3) complex combinational soil body, and (4) very complex combinational soil body (Figure 3.2, this volume). Fridland (1976a) used (1) elementary soil areal (ESA), (2) monocombinational soil areal (MSA), (3) and (4) polycombinational soil areal (PSA).

Closed soil bodies. Those soil bodies that occupy landforms with internal drainage, such as sinks in karst terrain. Contrast with *Open soil bodies* (Ruhe, 1969; Fridland, 1976a, p.52).

Coefficient of Dissection (CD). The ratio of the length of the boundary of a soil body to the circumference of a circle having the same area.

Classes	
Monolithic	= 1.0 – 1.5
Slightly dissected	– 1.5 – 2.5
Moderately dissected	= 2.5 – 5.0
Highly dissected	= > 5.0

See also *Soil body shape index.* (Fridland, 1976a, p. 37)

Combinational soil body. A more complicated unit of the soil cover than the elementary soil body. A combinational soil body is a cluster of elementary and/or combinational soil bodies. Some genetic relationship usually exists between components of a combinational soil body. See *Soil combination, Classificational ranks of soil cover, Soil Cover,* and Figures 3.2 and 3.7, in this book.

Composite mapping unit. A generalized mapping unit composed of several more detailed mapping units.

Combine. A soil combination (soilscape) with high pedologic contrast and mesorelief (10 to 100 m) locally and with large ESA's (1 ha to tens of ha in area), large enough for each ESA to be farmed separately. The genesis of the terrain preceded the genesis of the soil bodies. Interaction between soil bodies is unidirectional (Fridland, 1976a, p. 50).

Complex. A soil combination (soilscape) with high pedologic contrast and microrelief (less than 10 m) and small ESA's. Patterns of soil conditions spatially are 5 to 30 m across. ESA's are too small to be farmed separately. Genesis of the terrain and the soils were simultaneous (whereas in combines the terrain formed first, then the ESAs). Subsidence depressions may develop as the soilscape differentiates. Degree of bonding between components diminishes as the water table gradually lowers (Fridland, 1976a, p. 50).

Complexity of Soil Combinations. Variation within a soilscape of size, shape, interrelationships of consistituent soil bodies; number of soil boundaries per unit length of transect; degree of soil cover differentiation (Fridland, 1976a, p. 58).

Component. A constituent of a soil combination. For example: (1) sod podzolic ESA's may lie adjacent to (2) transitional ESAs and they adjacent to (3) bog ESAs. The first and third components (the constructional ones) may have large ESAs and the second (transitional) component may have small ESAs. Ratios between components may be calculated in terms of relative areas, sizes of bodies, degree of dissection of bodies (Fridland, 1976a).

Composition of Soil Combinations. List of soil species present (pedota) and count of number of species per unit area (such as 1 sq km). Proportionate extents of species listed (Fridland, 1976a).

Conscociation. "Mapping units in which only one kind of soil (or miscellaneous area) dominates each delineation to the extent that three-fourths or more of the polypedons fit within the taxon that provides the name for the mapping unit, or within similar soils. The most extensive kind of soil must fall within the range of the phase of the taxon ." (Soil Survey Staff, 1975b). Peterson (1981) explains that this term has been used by plant ecologists to identify stands of single species as opposed to associations of several plant species. Soil surveyors have applied it to soil map units in which only one identified soil component (plus allowable inclusions) occurs in each delineation.

Constructional components of soil cover. These are the end members of the pedologic spectrum. For example, in a terrain with well drained Spodosols on rises, and Histosols in adjacent depressions, the two named components are constructional components and intermediate soils are transitional components (Fridland, 1976a).

Continuous Universe. All the objects and classes under consideration which are not mutually exclusive, or which may be arbitrarily defined and not otherwise (such units may overlap). Contrast with *Particulate universe* (Knox, 1965).

Contrast in Soil Combinations. Degree of diversity or dissimilarity among components expressed in terms of taxonomy, or (preferably) in terms of one or a few diagnostic soil properties (Fridland, 1976a) such as texture, moisture regime, degree of erosion, extent of soil development (degree of podzolization; thickness of peat cover) or rating for crop production. If the elementary soil bodies of the dominant soil occupy less than 65% of the area, then the combination is contrastive. See *Heterogeneity.*

Creep Slope. A convex (in cross-section) shoulder slope. Troeh (1964) recognized two kinds: water-gathering (the slope is concave in plan view), and water-spreading (the slope is convex in plan view).

Davisian Erosion Cycle and the Evolution of Soilscapes. See Evolution of soil combinations.

Differentiation of soil combinations. The process or degree of subdivision of an area into progressively smaller soil bodies; or the difference between two soil combinations in terms of numbers of soil bodies per unit area.

The following spectrum is suggested:

Degree of Differentiation	Proportionate Area Occupied by the Background or Original Soil Body
Very Slight	90–100%
Slight	75–90
Moderate	60–75
Moderate to Strong	45–60
Strong	25–45
Very Strong	0–25

Fridland (1976) proposed a coefficient of classificational differentiation of soil cover (CDSC), a numerical index of the degree of contrast between components of a soil combination:

$$\mathrm{CDSC} = \frac{\sum_{n}^{i=1} E_i}{m} \cdot \frac{1}{n}$$

in which n is the number of taxonomic levels, m is the total number of soil units, and E_i is the number of soil units on every taxonomic level (Fridland, 1976a, p. 57).

Dissection. Degree of tortuosity of boundaries of an elementary soil body (ESB). The coefficient of dissection (CD) decreases as the size of the soil body increases, even without a change in shape. $\mathrm{CD} = S/[3.54 \times \sqrt{A}]$, where S is the boundary length, and A is the area of the soil body (Fridland, 1976a, p. 37).

Divide. A main interfluve between valleys (Ruhe, 1969).

Dynamics (cyclic) of a soil combination. The circulation (one-way or two-way) between contrasting elementary soil bodies, of water, of nutrients, and even particles. Each cycle of circulation may accomplish an infinitesimal increment of evolution of the degree of contrast in the pattern (Fridland, 1976a, p. 46). See *Evolution*, and *Stability of soil cover pattern.*

Ecosystem. An abiotic environment and assemblage of organisms within it (Forman and Godron, 1981).

Elementary soil areal (ESA). The simplest soil cover element. A soil formation free from any internal pedogeographical boundaries; its size is variable. A kind of soil occupying space that is bounded on all sides by other ESAs or non-soil formations. (Fridland, 1976a, p. 18) Three kinds of ESAs are recognized: (1) homogeneous ESAs; (2) sporadically patchy ESAs, which are interrupted by Limiting Pattern Elements (LPEs), patches of biological origin that contrast little with the background soil;

and (3) regular-cyclic ESAs, with a continuous network of polygons, each of which shows the full range of variation of one soil taxon that is found in the entire ESA. Note that regular cyclic patterns of contrasting soils do not themselves qualify as ESAs because they contain pedogeographical boundaries (Fridland, 1976a, pp. 19–21). See *Elementary soil body.*

Elementary soil body. A soil body without internal pedogeographic boundaries at the lowest taxonomic categorical level (below the soil series of the U.S.D.A.). Three kinds of elementary soil bodies were distinguished by Fridland (1976a) in his definition of Elementary Soil Areal (see above): (1) homogeneous elementary soil bodies; (2) slightly unhomogeneous elementary soil bodies that are biogenetically spotted (such as by digging activities of animals or by tree-tipping) with emphemeral (see below) patches of soil that have such low pedologic contrast to the matrix of the soil body that pedogeographic boundaries are not recognizable; (3) regular cyclic elementary soil bodies having so little pedologic contrast that pedogeographic boundaries are not recognizable. Sizes of elementary soil bodies range from about 0.5 sq m to thousands of sq km (see *size classification of soil bodies*). Note that the above definitions are for continuous elementary soil bodies. There are also punctured elementary soil bodies, the holes in which are occupied by smaller soil bodies or by not-soil. See *Combinational soil body.*

Evolution of soil cover combinations. Successive transformations through time of soil combinations beyond the scope of a single cycle of maintenance. An accumulative effect over a long period of time of the dynamics of a soil combination (which see). Relict elements survive in the presence of new elements. As conditions under and near the glacier waned, dewatering of the landscape took place. In places Tundra soilscapes evolved to desert ones. The Davisian cycle takes soil combinations from complexes to combines and back again. Peat wetlands may enlarge as a result of formation of ortstein on sides of depressions. Evolution of soil cover is slowest on peneplains in intertropical regions. Glaciated regions are subject to interruption of evolution of combinations by readvances of glaciers. In deserts deflation can locally make drastic changes in soils within a few years (Fridland, 1976, p. 97).

Evolution of soil cover pattern. Gradually changing nature of a soil combination by *Autometamorphosis* or *Parametamorphosis* (Fridland, 1976a, p. 97).

Exhumed soil body. A soil body, which, after formation, was buried, and now has been exhumed by removal of overburden, normally by natural erosion.

Flushing regime. A moisture regime in a body of soil in which an hydrologic connection exists between moisture in the solum and the water table. In areas of water deficit upward movement of water in the evapotranspiration stream may translocate salts to crests of low elevations (Vysotskii; in Rode, 1962). See *Non-flushing regime.*

Footslope position. In cross-sectional view of a slope profile, a colluvial concave hillslope position, above the alluvial toeslope and below the backslope (Ruhe, 1969).

Genesis of soil combinations. Genesis not only of soil pedons but of the entire soil cover, and of interrelationships among components. Modern and relict soil bodies may exist side by side. Genetic factors include relief, erosion, deposition, cryogenic and snow phenomena, non-uniformities (pedological and geological) in the soilscape, groundwater, patterns of plants and animals, anthropic patterns. Time is an indirect factor. Macroclimate is a homogenizing factor (Fridland. 1976, p.97).

Genetic unit. Area of the earth's surface that has been exposed to essentially uniform action of the soil forming factors, and therefore displays a degree of uniformity with respect to origin.

Generalized soil body. A generalized delineation on a soil map, accomplished by appropriate omission or smoothing of soil boundaries as recorded on a more detailed soil map.

Génon. "Volume of soil comprising all the pedons possessing the same structure, the same characteristics and resulting from the same pedogenesis. The *génon* is a *mappon:* it is a cartographic unit." (Boulaine, 1980, p. 102) Several kinds of *génons* are defined by Boulaine: (1) *génon simple:* a *mappon* representing a polypedon consisting of pedons 1 sq m in area. Such a *génon* may be homogeneous (pure) or may contain inclusions of contrasting pedons (spotted, or "*mâtiné*". (2) *génon variant:* A *mappon* representing a polypedon consisting of pedons up to 10 sq m in area. (3) *génon complex:* This is a *mappon* representing areals larger than 10 sq m in area. A *génon* body may be as large as 350 m in diameter. A description of a *génon* includes patterns present: striped, cellular, tongued, ruptic, mosaic, tachete, oriented, and gradient.

Geochemical aspect of genesis of soil combinations.

1. Type of weathering; type of transformation of humus; kinds of substances in eluvial members of catenas.
2. Acid-alkali and redox conditions; dilution of solutions.
3. Nature and sequence of geochemical barriers (acid-alkaline, redox, evaporative, temperature, biological, etc.) in the geochemically subordinate members of catenas (transaccumulative supersaturated ones).
4. Geochemical history of the landscape related to evolution of relief and succession of climates (Fridland, 1976a, p. 93).

Geomorphology. The systmatic examination of landforms and their interpretation as records of geologic history (Howell, 1957).

Graphic generalization. Simplification of a map pattern by smoothing of lines.

Griva. Russian term for gently sloping low narrow ridge only a few meters high and extending for as much as 10 km (Fridland, 1976a, p. 74).

Hand specimen. Sample of soil of a size that can be conveniently held in the hand. If such a sample is taken as representative of larger volumes of soil, then in effect it becomes a member-body in a soil universe that defines soil as isotropic (homogeneous). This definition may be useful to engineers but is not adequate for the purposes of pedologists (Knox, 1965).

Head slope. A hillslope, as seen in plan view, with concave boundaries above and below, situated in a hollow between interfluves (Ruhe, 1969).

Heterogeneity of a soil combination. An index of combined contrast (see above) and complexity (see above) of a soil combination.

Heteronomous soil. Soil influenced not only by atmospheric moisture, but also by additional influx of moisture from other areas (Fridland, 1976, p. 85). See *Autonomous soil.*

Homologous series of soil combinations. A set of binary or tertiary soil combinations that exhibit the same pattern, relationship and genetic trend, but have different pedologic content. Examples are hillock-depression pairs, with the first-named soil representing the hillock position: Chernozem-Solonetz; Chestnut-Solonetz; Brown semidesert soil-Solonetz; Gray-brown desert soil-Solonetz; Podzol soil-Bog podzolic soil (Fridland, 1976, pp. 89–91). See Analogous soil combinations.

Horizon. See *Soil Horizon.*

Human effects on soil combinations and soil cover patterns.

1. Increases in contrast by (a) acceleration of erosion, (b) intensification of salinity, (c) reduction in transpiration, resulting in increased surficial gleyzation.
2. Increase in number of soil bodies (classed as "mosaics" by Fridland) by the construction and abandonment of shelters, corrals, wells.
3. Decrease in the contrast in areas subjected to (a) irrigation, (b) drainage, and (c) leveling and cultivation.
4. Impairment or enhancement of soil fertility (Fridland, 1976a, p. 111).

Ilovka. Russian term for a silty clay mixed with muck, rendering the soil humic (Fridland, 1976a, p.8).

Inclusion. An "impurity" or unnamed foreign soil present within an area delineated and labeled as a certain mapping unit. The reader of the map is not explicitly informed of the presence or identity of these soils, and his ability to use the map as a predictive tool is reduced.

Individual. A thing complete in itself (Cline, 1949). A soil individual is a complete soil. See *Elementary soil body.*

Intensity of a soil map. The number of delineations demarcated per unit area on a soil map (Laker, 1972).

Initial material of soil. "The unconsolidated and more or less chemically weathered mineral or organic matter from which the solum of soils is developed by pedogenic processes." (Soil Science Society of America, 1979) The term "initial material" is free of the anthromorphic connotation carried by the term "parent material."

Interfluve. A subsidiary spur extending from a main divide between tributary valleys. First, second, etc. orders of interfluves are recognized (Ruhe, 1969).

Isotropic. A condition of a volume of an ideal initial soil material of being homogeneous. See *Anisotropic.*

Landscape. "A kilometer-wide area where a cluster of interacting stands or ecosystems is repeated in similar form." (Forman and Godron, 1981) "A stretch of country as seen from a particular vantage point." (Harris, 1968).

Landscape positions of soil bodies, as seen in cross-section. These are: summit, shoulder, backslope, footslope, toeslope (Ruhe, 1969).

Landscape positions of soil bodies as seen in plan view. These are: nose slope, side slope, head slope, divide, and interfluve (Ruhe, 1969).

Landsurface catena. Toposequence of landscape units in full three dimensions (Conacher and Dalrymple, 1977).

Landscape unit. See *Soil landscape unit.*

Liman. Russian term for a shallow depression in steppe that is periodically wetted by snow melt, ranging from a few dozen sq m to a few sq km in area. Pools of open water are present in the spring, but they dry up by summer (Fridland, 1976a, p. 200).

Limiting pattern element (LPE). Small areas measuring only a few square meters or tens of square meters in area and distinguished by the occurrence of soils contrasting with surrounding soils as a result of biotic influence on the LPE. Tree tipping, rotting of tree stumps, and animal burrowing are examples of factors that create LPEs. Under the canopy of an isolated tree in a savanna, the soil has a lower content of organic matter than does the soil of the open savanna. The spot under the tree is the LPE. LPEs differ from ESAs in respect to size, origin, and pattern of occurrence on the landscape (Fridland, 1976a, p.20).

Limiting structural element (LSE). See *Limiting pattern element.*

Mapping unit. A legend item on a map, defined to correspond to a certain soil, or set of soils. Usually limitations of map scale in relation to landscape complexity mean that mapping units are composed of several kinds of soil, not all of which are actually specified in the legend. See inclusion. Contrast with *Taxonomic unit, Genetic unit, Landscape unit.*

Mappon. "Abstract soil, delineated on a soil map and defined in terms of a reference soil taxon" (Boulaine, 1980).

Massive. A soil structure classification in which the soil particles exhibit no aggregation into peds. Usually found in soils of medium texture.

Member-body. A body in a physical universe that qualifies for membership in a class (Knox, 1965).

Merging of elementary soil bodies. By processes of weathering, mass wasting, collapse, and erosion, two adjacent sinks or dolines may merge, uniting two formerly separate elementary soil bodies into a single larger one. The increase in size is not accompanied by a qualitative change (Fridland, 1976a).

Minimal soil combination. This is the unit cell of a particular kind of soil cover. It is a cluster of different kinds of elementary soil bodies that is inclusive enough to represent the principal pedologic composition and arrangement characteristics of an extensive soil combination containing many repetitions of the unit cell. Thus, the essential characteristics of a monocombinational soil body can be concisely represented by such a unit cell, or minimal soil composition (Fridland, 1976a).

Monocombinational soil body. See *Soil combination.*

Monofactorial soil combination. A soil combination, whose pattern is dominated by a single factor. For example, river flooding dominates the soil pattern of many Fluvents (alluvial soils) (Fridland, 1976a, p. 93).

Mosaic. A soil combination that is usually lithogenetic and exhibits high contrast of component soil bodies, between which little or no pedogenic interaction takes place (Fridland, 1976a).

Multifactorial soil combination. A soil combination that shows the influence of two or more factors. For example, a level body of solonetzous Chestnut soil is interrupted by (1) mounds occupied by solonchak Solonetz, and (2) Zapadina depressions occupied by meadow Chestnut soils. The two factors at work here are: (1) animal activity at the mounds where faunal activity has brought salt to the surface, and (2) subsidence at the sites of the depressions (Fridland, 1976a, p. 93).

Natural soil individual. A real soil body, whether affected by human activity or not, that may be observed in a terrain (Knox, 1965). See *Artificial soil individual.*

Non-flushing regime. A moisture regime in a soil body in which a "dead" zone separates moisture in the solum from the water table; that is, there is no evidence that water percolates down or moves upwards through that zone. The water films in the "dead" zone are so thin that movement of water is so slow as to be considered inconsequential (Vysotskii; in Rode, 1962).

Nose slope. A hillslope, as seen in plan view, with convex boundaries above and below at the exposed end of the interfluve (Ruhe, 1969, 1974).

Not-soil body. A volume of not-soil (see *Soil* below), such as open water, glacial ice, flowing hot lava, large mass of salt, gypsum, massive rock, or even unconsolidated sediments, provided such materials have not been influenced by soil genetic factors to produce a soil body that differs from the initial material body.

Open soil body. A soil body that has surface drainage connected to stream channels of regional drainage networks (Ruhe, 1969). See *Closed soil body.*

Padina. Russian term for flat, saucer-like depression in a steppe, possibly larger than a *Zapadina* (Fridland, 1976a, p.64).

Parametamorphosis. A change in soil cover caused by an external factor, such as regional drop in water table, or regional dissection by an extending stream network (Fridland, 1976a, p. 104). See *Autometamorphosis.*

Parent material of soil. See *Initial material.*

Particulate universe. All of the objects and classes under consideration consist of countable non-overlapping bodies (Knox, 1965). Contrast with *Continuous universe.*

Patchwork (Patchiness). A soil combination of low relief and low pedologic contrast. Examples: (1) Patchwork of medium and strongly podzolized soils; (2) patchwork of typical and leached Chernozems (Fridland, 1976a, p. 42).

Ped. "A unit of soil structure, such as an aggregate, crumb, prism, block, or granule, formed by natural processes (in contrast with a clod, which is formed artificially)." (Soil Science Society of America, 1979) Volumes of soil that are structureless (massive, or single-grain) are called *apedal.*

Pedal. The property of being composed of peds. Massive and single-grain soils that lack pedal structure are said to be *apedal.*

Pedochore. A combinational soil body (Hasse, 1968).

Pedology. The science that addresses soils, their properties, origins, and occurrence on the landscape. Soil science.

Pedon. "A three-dimensional body of soil with lateral dimensions large enough to permit the study of horizon shapes and relations. Its area ranges from 1 to 10 square meters. Where horizons are intermittent or cyclic and recur at linear intervals of 2 to 7 meters, the pedon includes one-half of the cycle. Where the cycle is less than 2 meters, or all horizons are continuous and of uniform thickness, the pedon has an area of approximately 1 square meter. If the horizons are cyclic, but recur at intervals greater than 7 meters, the pedon reverts to the 1 square meter size, and more than one soil will usually be represented in each cycle" (Soil Science Society of America, 1979) (see also: Johnson, 1963; Arnold, 1964; Soil Survey Staff, 1975a).

Pedota. List of soil species of a defined region (Hole, 1976).

Pedotop. A natural volume of soil on the landscape that may be homogeneous or complex in respect to membership and pattern. It may be classified as (1) homogeneous (monomorphic), (2) semipolymorphic complex, or (3) polymorphic complex (Hasse, 1968).

Phénon. "Objective soil, real, concrete, in place (pedon and soil units)" (Boulaine, 1980).

Plowland type. Term used by Sibertsev for "topographic unit" (he identified 22 such units) that exhibits recurrent soil combinations "limited within definite schemes." Relief is the principal factor that determines soil cover pattern. This term is a synonym for major soil combination (Fridland, 1976a, p.8).

Pocosin "A swamp, usually containing organic soil, and partly or completely enclosed by a sandy rim. The Carolina Bays of the Eastern United States" (Soil Science Society of America, 1979).

Podburs. Russian term for a group of non-gleyed, non-podzolized soils that are widespread in cold regions of Eurasia. The term is now applied to a particular soil type that is analogous to Subartic Brown Forest soils of North America (Fridland, 1976a, p. 64).

Pody. Russian term for a large, shallow, flat depression (oval or elongated) in steppe, measuring 7 to 8 km long and half as wide. These are subsidence depressions of distintegrating rock (Fridland, 1976a, p. 39).

Polder. A tract of low land, reclaimed from the sea or other body of water by dikes, dams, etc., in which the water level in ditches is artificially maintained by engineering works (Van Hessen, 1970).

Polesie. Russian term for an extensive tract that consists mainly of forested wetlands (the word is derived from "les" meaning forest). The term is also used as the proper name for such tracts in the northwestern part of the Ukraine, Belorussia, and Poland (Fridland, 1976a, p. 165).

Polycombinational soil body. See *Classificational rank of the soil cover.*

Polypedon. A group of contiguous similar pedons. A polypedon has a minimum area of 1 sq m, and an unspecified maximum area. (Johnson, 1963, p. 215; Soil Survey Staff, 1975) Every polypedon can be classified into a soil series, but a series normally has a wider range that that shown by a single polypedon. The polypedon is more complex than an elementary soil body and is primarily an expression of a taxonomic definition. Therefore, a polypedon is not a definite, discrete, individual.

Primary soil particles. Natural individuals in a particulate universe. These do not constitute a complete soil. They include particles of primary minerals (igneous in origin), secondary minerals (metamorphic and sedimentary in origin), and organic particles (Knox, 1965).

Processes of soil formation. Elementary soil processes are those which operate in the framework of factors of soil formation to form soils.

Punctate patchwork. This is a combinational soil body that is punctured, with elementary soil bodies (and/or bodies of not-soil) set in the lacunae.

Punctate soil body. An elementary soil body that is punctured and has smaller elementary soil bodies (or bodies of not-soil) inset in the lacunae. See *punctate patchwork.*

Relief. The difference in elevation between the high and low points of a land surface. Fridland (1976a) proposed the following categories: (i) microrelief, 0 to 10 m locally; (ii) mesorelief, 10 to 100 m locally; and (iii) macrorelief, more than 100 m locally. Hole (1978) proposed (i) very low: 0 to 1 m; (ii) low: 1 to 10 m; (iii) moderate: 10 to 100 m; (iv) high: 100 to 1,000 m, and (v) very high: $>$ 1,000 m. We note that the scope of the word "local" has not been defined.

Saz. Russian term for habitat with high and permanent water table in arid regions of central Asia (Fridland, 1976a, p. 215).

Sculpturing. External appearance of a pattern on the surface of a pollen grain. Classes of sculpturing (accidental, reticular, striated, pubescent) have analogues in microrelief of soil cover (See Faegri, Iverson, Waterbolk, 1964).

Shoulder position. This is the convex hillslope position ("waxing slope") below the summit and above the backslope (Ruhe, 1969).

Side slope. A hillslope, as seen in plan view, with fairly straight boundaries above and below, situated on the side of an interfluve (Ruhe, 1969, 1974).

Simple soil combination. A soil combination composed of elementary soil bodies. Fridland (1976a, p. 42) classified simple soil combinations as follows:

Simple Soil Combinations	Soil Catenas (soils genetically related)	Complex	High	Micro
		Patchiness	Low	Micro
		Combine	High	Meso
		Variation	Low	Meso
	Soil Sequences (soils not genetically related)	Mosaic	High	Meso
		Tachets	Low	Meso

Single-grain structure. (Obsolete) A soil structure classification in which the soil particles occur almost completely as individual or primary particles with essentially no secondary particles or aggregates present. (Usually found only in extremely coarse-textured soils.) (Soil Science Society of America, 1979).

Size classification of soil bodies. A possible classification, combining some information from Fridland (1976a), and from Hole (1978, and Figure 4.4, this volume) for elementary soil bodies is as follows:

Very Large (megamassive) > 100 ha (1 sq km)	Very Coarse (100 to 10,000 sq km)
	Coarse (1 to 10 sq km)
Large (large massive) 10 to 100 ha	Intermediate
Medium (medium massive) 5 to 10 ha	Mediate, large
Small (small massive) 1 to 5 ha	Mediate, small
Very Small (micromassive) < 1 ha	Fine

Sociations. Soil combinations in a hierarchial classification by size of their bodies (Simonson, 1971).

Soil. "(i) The unconsolidated mineral *and organic** material on the immediate surface of the *land* that serves as a natural medium for the growth of land plants, *or that responds to diurnal and seasonal climatic and microclimatic conditions in the absence of plants (as in parts of Anarctica)**. (ii) The unconsolidated mineral *and organic matter** on the surface of the *land** that has been subjected to and influenced by genetic and environmental factors of: *parent material, climate* (including moisture and temperature effects), *macro- and microorganisms,* topography, all acting over a period of *time* and producing a product–soil–that differs from the material from which it is derived in many physical chemical, biological and morphological properties, and characteristics." (Soil Science Society of America, 1979, except for starred (*) words and phrases).

Soil areal (SA). A soil body on the landscape, including elementary soil areals (ESAs), and soil combinations. Limiting pattern or structural elements (LPEs, LSEs) are considered ephemeral and hence do not attain the status of soil areals (Fridland, 1976a, p. 15).

Soil association. "(i) a group of defined and named taxonomic soil units occurring together in an individual and characteristic pattern over a geographic region., comparable to plant associations in many ways. (Sometimes called 'natural land types.') (ii) A mapping unit used on general soil maps, in which two or more defined taxonomic units occurring together in a characteristic pattern are combined because the scale of the map or the purpose for which it is being made does not require delineation of the individual soils." (Soil Science Society of America, 1979) See also Simonson, 1971, p. 960).

Soil autecology. The study of environmental relationships of bodies of a particular species of soil. Contrast with *Soil synecology.*

Soil body. (i) A volume of complete soil in the landscape. (ii) A natural soil individual (as distinct from an artificial one). (iii) A volume of a remnant of a formerly complete soil in the landscape. See *elementary soil body, monocombinational soil body, polycombinational soil body, generalized soil body, buried soil body, exhumed soil body, polypedon, soilscape, pedochore.* Simonson (1978) defines soil bodies as open, three-phase (i.e., solid, liquid, gaseous) systems, consisting of the same elements and components, but in different proportions from place to place, and to which and from which substances and energy are added and lost as soil horizon differentiation proceeds. See also Butler (1983) for yet another perspective.

Soil body shape index. Index of the irregularity of the boundary of a soil body, determined by dividing the length of the boundary by the perimeter of a circle of the same area as the soil body (Hole, 1978). See *Coefficient of dissection.*

Soil boundary. A line, or band, composed of points on the landscape at which soil characteristics are transitional between those of adjacent soil bodies, within which gradients of change are spatially less rapid than at the boundary. Fridland (1976a) noted three kinds of soil boundaries: (1) sharp (abrupt on the land); (2) distinct (neither abrupt nor gradual); and (3) gradual (the cartographer usually draws a line in the middle of the zone of transition without indicating the true width of the boundary). Proportionate lengths of the three kinds of boundaries may be reported for a given soil body. Quantification of distinctness of soil boundaries is suggested (based upon comments by Fridland, 1976a) as follows: Very sharp, $<$ 0.3 m; sharp. 0.3 to 3 m; distinct, 3 to 5 m; gradual, 5 to 10 m; diffuse, $>$ 10 m. Very broad diffuse boundaries are commonly subdivided arbitrarily into soil body delineations on maps. Note that the form of a boundary may vary depending upon the specific property, or properties, considered. (See Campbell, 1977).

Soil classification. "The systematic arrangement of soils into groups or categories on the basis of their characteristics." (Soil Science Society of America, 1979).

Soil combination (SC). (Fridland, 1976a, p. 42) Soil combinations are soil cover units that are more complex than are elementary soil areals (ESAs). A simple or monocombinational soil areal (MSA) is composed of coterminous alternating ESAs that are somewhat genetically related, with or without non-soil areals. Polycombinational soil areals (PSAs) are complicated combinations composed of MSAs with or without additional ESAs and nonsoil areals. Microcombinations (micro-MSAs) consist of recurrent successions of fine or micromassive ESAs (diameters are commonly 1 to 2 meters but may be as large as 50 m) that are related to microrelief (local relief is mostly < 2 m but may exceed 10 m). The ESAs are too small to be farmed separately. The terms "complex" and "patchiness" designate two kinds of microcombinations and are contrasting and noncontrasting, respectively. Mesocombinations are composed of large ESAs (10 to 100 ha) that can be farmed separately, or of such ESAs together with microcombinations, all on terrain of mesorelief (10 to 100 m). The terms "combine" and "variation" designate two kinds of mesocombinations, constrasting and noncontrasting respectively. Macrocombinations are those of hills,, mountains, and plains with macrorelief (more than 100 m locally). Sibertsev used the terms "soil combination," "soil type," and "plowland" in much the same sense. See *Combinational soil body.*

Soil complex. A mapping unit composed of soil units occurring in a pattern too complex to be individually represented at the scale of the survey (Nygard and Hole, 1949).

Soil component ratio. Ratio between components of a soil combination. Ratios based upon proportionate areas of components are commonly reported. These ratios may be constant through the years, or, as in the case of alluvial soils, may change from season to season (Fridland, 1976a, p. 58).

Soil continuum. The soil cover or blanket on the land surface, interrupted though it is by bodies of not-soil (which see).

Soil cover. "The entirety of soils occurring on a territory." (Fridland, 1976a, p. 15) A three-dimensional body with horizontal and vertical extents respectively equal to the area and depth of the soils on a territory.

Soil cover differentiation. See *Differentiation of soil combination.*

Soil cover pattern. This is the spatial arrangement of soil areals (bodies), and has been called "soil cover structure." Its description requires not only specification of the arrangement of soil bodies, but also the list of taxonomic components, their proportionate extents, sizes, shapes, and number of soil bodies per unit area, distinctness of boundaries, degree of pedologic contrast between neighboring bodies, interrelationships between bodies and evolution of the pattern (Fridland, 1976a, p. 6).

Cline (1972) considered soil cover pattern from a utilitarian agronomic point of view and distinguished simple and complex soil patterns, which he defined as follows: (1) Simple soil pattern: "At a scale of 1:20,000 delineations representative of areas having similar use potential would be predominately larger than 1000 ha, and those representing areas having similar requirements for management systems of intensive uses such as crops would be predominately larger than 10 ha." (2) Complex soil pattern: "At a scale of 1:20,000 delineations of similar use potential would be predominately smaller than 1000 ha or those of similar needs for mangement systems would be smaller than 10 ha or both."

Soil genesis. "(i) The mode of origin of the soil with special reference to the processes or soil-forming factors responsible for the development of the solum, or true soil, from the unconsolidated material. (ii) A division of soil science concerned with soil genesis (i)." (Soil Science Society of America, 1979) Soil genesis is the continuing response of a complex body of mineral and organic materials to a never-ending sequence of impacts and overprints by climatic and organic events and attendent physical, chemical, and biotic processes.

Soil geography. "A subspecialization of physical geography concerned with the areal distributions of soil types." (Soil Science Society of America, 1979).

Soil horizon. "A layer of soil or soil material approximately parallel to the land surface and differing from adjacent genetically related layers in physical, chemical, and biological properties or chacteristics, such as structure, texture, consistency, kinds and numbers of organisms present, degree of acidity or alkalinity, etc." (Soil Science Society of America, 1979).

Soil individual. A complete soil. See *Elementary soil body.* Natural soil individuals are distinguished from artificial ones.

Soil landscape. The soils portion of the landscape. *Soilscape* is an abbreviation.

Soil landscape unit. A spatial aggregate of soil individuals with lateral boundaries determined by the geographic pattern of change in soil characteristics according to objective boundary criteria (Cline, 1949; Knox, 1965). Usually corresponds to a *Genetic unit.* Compare with *Mapping unit* and *Taxonomic unit.*

Soil phase. "A subdivision of a soil type or other unit of classification having characteristics that affect the use and management of the soil but which do not vary sufficiently to differentiate it as a separate type. A variation in a property or characteristic such as degree of slope, degree of erosion, content of stones, etc." (Soil Science Society of America, 1979). See *Consociation.*

Soilscape. The pedologic portion of the landscape. In this book, "soilscape" is considered an abbreviation of the term "soil landscape." Soilscapes consist of the upper portion of the landscape plasma, which is the total mass of unconsolidated geologic and pedologic material present. (Buol, Hole, and McCracken, 1980). Hole's (1978) use of "soilscape" for a particular assemblage of soil bodies is not followed here. Rather, the term "combinational soil body" is preferred.

Soil series. "The basic unit of soil classification, being a subdivision of a family and consisting of soils that are essentially alike in all major profile characteristics except the texture of the A horizon. The soil series is now the lowest category in the natural classification system or in soil taxonomy. Phases of soil series are now the major components shown in detailed soil maps in the USA." (Soil Science Society of America, 1979)

Soil space. "The soil horizons and their mineral and humus particles, also the gas- and water-filled pores, and all the organisms that inhabit it." (Jenny, 1980) See *Vert space.*

Soil synecology. The study of entire soil landscapes or soil combinations as pedological communities. Contrast with *Soil autecology.*

Soil type. "Obsolete; see Soil series. Formerly in the U.S. soil classifications prior to the publication of Soil Taxonomy (1975a): (i) The lowest unit in the natural system of soil classification; a subdivision of a soil series and consisting of or describing soils that are alike in all characteristics including the texture of the A horizon or plow layer. (ii) In Europe, roughly equivalent to a great soil group." (Soil Science Society of America, 1979).

Soil variant. "A soil whose properties are believed to be sufficiently different from other known soils to justify a new series name but comprising such a limited geographic area that creation of a new series is not justified." (Soil Science Society of America, 1979).

Soil zone. This is an areal (body) of soil combinations consisting of autonomous soils of one or more kinds (differences among them being due to nonclimatic factors such as initial material, time,etc.), together with related subordinate soils. The territory of a soil zone may also contain soil combinations with autonomous soils that are characteristic of other zones, but play a subordinate part and occupy limited areas with specific conditions of soil genesis (Fridland, 1976a, p. 85).

Solum. (plural: *sola*) "The upper and most weathered part of the soil profile; the A and B horizons." (Soil Science Society of America, 1979).

Sor. Russian term for a type of salt lands in Soviet Central Asia (Fridland, 1976a, p. 177).

Stability of the soil cover pattern. Unchanging nature of the pattern of a soil combination as a result of dynamic, cyclic variation through the seasons. For example, runoff from elevated bodies of Spodosols into depressed bodies of Histosols is a dynamic process that maintains the pattern of a Spodosols-Histosols combination (Fridland, 1976a, p. 97). Contrast with *Evolution of the soil cover pattern.*

Summit position. Highest hillslope position, above the shoulder position (Ruhe, 1969).

Syrt. Russian term for a flat, clayey watershed upland (Fridland, 1976a, p. 68).

Spatial generalization. Simplification of a map pattern by combining those mapping units that tend to consistently adjoin one another on the detailed map, without regard to their degree of pedologic similarity.

Sporadic-Patchy Elementary Soil Areals (ESAs). A group of elementary soil bodies defined by Fridland (1976a) as having sporadic patches of limiting pattern elements on a homogeneous background. For example, under the tall grasses of savannas homogeneous backgound soil has developed, and under canopies of trees scattered in the savannas are limiting pattern elements which consist of patches of soil with relatively low content of organic matter.

System. A complex of interrelated elements displaying "structural similarity and isomorphism in different areas" (Bertelanffy, 1956). Soil cover patterns with strong dynamic bonds between components are undeniably systems. Patterns with tenuous bonds are conditional systems. Open soil cover patterns exchange both energy and materials with ambient environment. Closed soil cover patterns exchange only energy with ambient environment (Fridland, 1976a, p. 4).

Tachet. A soil combination body that consists of limiting pattern or structural elements, which are biologically produced spots of low contrast (Fridland, 1976a, p. 42).

Tachety. Spotted nature of a soil cover pattern consisting of tachets (Fridland, 1976a, p. 42).

Talud. A colluvial soil body formed at the down-slope margin of a cultivated field against a manmade barrier following the contour (a stone wall, hedge, or fence row of brush) (Soil Survey Staff, 1951, p. 252).

Taxon. A unit of soil classification; an ideal soil, defined on the basis of a small number of hierarchial characteristics, chosen because of their natural significance and their usefulness in discriminating between soils (Boulaine, 1980).

Taxonomic generalization. Simplification of a map pattern by combining those mapping units on the detailed map that are most nearly similar to one another in respect to pedologic characteristics.

Taxonomic unit. A subdivision within a taxonomic system. For example, the Mollisols consitute a taxonmic unit. Contrast with *genetic unit* and *mapping unit.*

Tectonic and historical geological factors. Factors of the nature indicated that influence the relations among soil cover components, the geometric features, and the evolutionary trends of combinations (Fridland, 1976a, p. 92).

Terrain. A landscape, or "lay of the land." (Bloom, 1978).

Terrane. An area of a certain structure or rock type, such as "granitic terrane," or "limestone terrane." (Bloom, 1978).

Tessera. "The operational unit which we collect in the field, examine, and analyse." It is usually smaller than a pedon, but larger than a hand specimen. An eco-tessera samples both soil and vegetation. A soil tessera samples only the soil part of the ecosystem (Jenny, 1965).

Time-zero. The moment at which soil formation begins and soil profiles and soil bodies and arrangements of them are initiated. The sudden draining of a lake, with subsequent exposure of the bottom to impact of climate and organisms, illustrates how time-zero may be introduced into an area. Theoretically each body of soil had a time-zero. Abrupt changes in environmental conditions may superimpose upon a soil body a succession of time-zeros so that the body is polygenetic.

Toeslope position. See *Alluvial toeslope.*

Toposequence. See *Chorosequence.*

Transitional component. A soil body in a transitional belt between constructional components in a soil cover.

Universe. A superclass containing all objects and classes under consideration. A particulate universe may be distinguished from a continuous universe (Knox, 1965).

Uval. Russian term for a ridge, usually of more than 200 m in local relief, with a broad, rounded crest, and an indistinct foot (Fridland, 1976a, p. 134).

Variation. A soil combination consisting of large ESAs of low contrast on a terrain of mesorelief (10 to 100 m locally). This forms a low contrast analogue of the combine (Fridland, 1976).

Vert Space. "All the above-ground parts of the land ecosystem, the plants, the animals, and the voids between them." (Jenny, 1980).

Vertical Zonality. Arrangement of climatic zones at different elevations on mountainsides (Fridland, 1976a).

Zapadina. A Russian term for a flat, shallow depression in steppe. The depression is wetter than is the surrounding terrain (Fridland, 1976a, p. 46).

Zonal-Provincial Factors. The factors that are zonal which influence the range of soil cover components. See *tectonic factors.*

Bibliography

Aandahl, A.R. 1972. *Soils of the Great Plains.* Lincoln, Nebraska: A.R. Aandahl.

Abmeyer, W., and H.V. Campbell. 1970. *Soil Survey of Shawnee County, Kansas.* Washington, D.C.: Soil Conservation Service, U.S.D.A.

Agterberg, F.P.. 1965. The Technique of Serial Correlation Applied to Continuous Series of Element Concentration Values in Homogeneous Rocks. *Journal of Geology.* Vol. 73, pp. 142–54.

Anderson, G.D., and D.G. Herlocker. 1973. Soil Factors Affecting the Distribution of the Vegetation Types and Their Utilization by Wild Animals in the Ngorongoro Crater, Tanzania. *Journal of Ecology.* Vol. 61, pp. 627–51.

Anderson, R.L. 1956. A Comparison of Discrete and Continuous Models in Agricultural Production Analysis. Pp. 39–61 in *Methodological Procedures in Economic Analysis of Fertilizer Use Data* (E.L. Baum, E.O. Heady, and J. Blackmore, eds.). Ames: Iowa State College Press.

Arnold, R.W. 1965. Cyclic Variations and the Pedon. *Soil Science Society of America Proceedings.* Vol. 28, pp. 801–4.

Arroues, Kerry D. 1982. Soils and Plants: A Way to Read the Landscape. *Soil Survey Horizons.* Vol. 23, No. 4, pp. 16–20.

Ball. D.F., and W.M. Williams. 1968. Variability of Soil Chemical Properties in Two Uncultivated Brown Earths. *Journal of Soil Science.* Vol. 19, pp. 379–91.

Bartelli, L. J. 1966. General Soil Maps—A Study of Landscapes. *Journal of Soil and Water Conservation.* Vol. 21, pp. 3–6.

Bartram, John. 1751. *Observations on the Inhabitants, Climate, Soil, Rivers, Productions, Animals, and Other Matters Worthy of Notice.* London.

______. 1791. *Travels Through North and South Carolina, Georgia, East and West Florida.* Philadelpia.

Basu, J.P., and P.L. Odell. 1974. Effects of Intraclass Correlation Among Training Samples on the Misclassification Probabilities of Bayes' Procedure. *Pattern Recognition.* Vol. 6, pp. 13–16.

Baumgartner, A., and E. Reichel. 1975. *The World Water Balance: Mean Annual Global, Continental and Maritime Precipitation, Evaporation, and Run-off.* English translation by Richard Lee. Amsterdam: Elsevier.

Baxter, F.P., and F.D. Hole. 1967. Ant *(Formica cinerea)* Pedoturbation in a Prairie Soil. *Soil Science Society of America Proceedings.* Vol. 31, pp. 425–28.

Beckett, P. 1977. Cartographic Generalization. *Cartographic Journal.* Vol. 14, pp. 49–50.

Beckett, P.H.T., and R. Webster. 1971. Soil Variability: A Review. *Soils and Fertilizers.* Vol. 34, pp. 1–15.

Beinroth, F.D. 1978. Opening Address. in *Proceedings, First International Soil Classification Workshop* (Camargo, M.N., and F.H. Bienroth, eds.). Washington, D.C.: Agency for International Development, Office of Agricultural Development Bureau.

Bertalanffy, L. 1956. General Systems Theory. *Yearbook of the Society for the Advancement of General Systems Theory.* Vol. 1. Ann Arbor, Michigan, U.S.A.

Bidwell, O.W., and C.W. McBee. 1973. *Soils of Kansas.* Kansas Agricultural Experiment Station, Manhattan.

Bilzi, A.F., and E.J. Ciolkosz. 1977. A Field Morphology Rating Scale for Evaluating Pedological Development. *Soil Science.* Vol. 124, pp. 45–48.

Bleeker, P., and J.G. Speight. 1978. Soil-Landform Relationships at Two Localities in Papua New Guinea. *Geoderma.* Vol. 21, pp. 183–98.

Bloom, A. 1978. *Geomorphology.* Englewood Cliffs: Prentice-Hall.

Bridges, E.M. 1977. Soil Geography, Its Content and Literature. *Journal of Geography in Higher Education.* Vol. 1, pp. 61–72.

______. 1978. Interaction of Soil and Mankind in Britain. *Journal of Soil Science.* Vol. 29, pp. 125–39.

Boulaine, J. 1969. Sol, Pédon, et Génon: Concepts et définitions. *Bulletin l'Association Francaise pour l'Etude du Sol.* No. 2, Paris. pp. 31–40.

______. 1975. *Géographie des sols.* Paris: Presses Universitaires de France.

______. 1980. *Pédologie Appliquée.* Paris: Masson.

Boulding, Kenneth E. 1956. *The Image; Knowledge and Life and Society.* Ann Arbor: University of Michigan Press.

______. 1980. Science: Our Common Heritage. *Science.* Vol. 207, pp. 831–36.

Bouma, J. 1974. New Concepts in Soil Interpretations for On-Site Disposal of Septic Tank Effluent. *Soil Science Society of America Proceedings.* Vol. 38, pp. 941–46.

Bouma, J., and F.D. Hole. 1971. Soil Structure and Hydraulic Conductivity of Adjacent Virgin and Cultivated Pedons of Two Sites: A Typic Argiudoll (silt loam) a Ruptic Eutrochrept (clay). *Soil Science Society of America Proceedings.* Vol. 35, pp. 316–19.

Bouma, J.; W.A. Ziebell; W.G. Walker; P.G. Olcott; E. McCoy; and F.D. Hole. 1972. *Soil Absorption of Septic Tank Effluent.* Information Circular No. 20. Madison: Univeristy of Wisconsin Extension; Geological and Natural History Survey.

Brewer, R. 1976. *Fabric and Mineral Analysis of Soils.* New York: Robert E. Krieger Publishing Co.

Bridges, E.M. 1977. Soil Geography, Its Content and Literature. *Journal of Geography in Higher Education.* Vol. 1, pp. 61–72.

______. 1978. Interaction of Soil and Mankind in Britain. *Journal of Soil Science.* Vol. 29, pp. 125–39.

Brophy, D.M. 1972. *Automated Linear Generalization in Thematic Cartography.* Unpublished Master's Thesis, University of Wisconsin, Madison.

Brownfield, Shelby H. 1976. *Soil Survey of Bartholomew County, Indiana.* Washington, D.C.: Soil Conservation Service, U.S.D.A.

Bucher, August Leopold. 1827. *Von der Hindernissen, welche der Einfuhrung eines besseren Ganges bym Vortrage der Erkunde auf Schulen im Wege stehen.* Coslin.

Bunge, William. 1962. *Theoretical Geography.* Lund Studies in Geography. Lund, Sweden: Gleerup.

Bunting, B. T. 1965. *The Geography of Soil.* London: Hutchinson.

Buol, S.W. 1976. Erosion and Taxonomy. *Soil Survey Horizons.* Vol. 17, No. 2, pp. 17–19.

Buol, S.W.; F.D. Hole; and R.J. McCracken. 1980. *Soil Genesis and Classification.* Iowa State University Press, Ames.

Burgess, T.M., and R. Webster. 1980a. Optimal Interpolation and Isarithmic Mapping of Soil Properties: I. The Semi-Variogram and Punctual Kriging. *Journal of Soil Science.* Vol. 31, pp. 315–31.

______. 1980b. Optimal Interpolation and Isarithmic Mapping of Soil Properties: II. Block Kriging. *Journal of Soil Science.* Vol. 31, pp. 333–41.

Burrough, P.A. 1983a. Multiscale Sources of Spatial Variation in Soil. I. The Application of Fractal Concepts to Nested Levels of Soil Variation. *Journal of Soil Science.* Vol. 34, pp. 577–97.

______. 1983b. Multiscale Sources of Spatial Variation in Soil. II. A Non-Brownian Fractal Model and Its Application. *Journal of Soil Science.* Vol. 34, pp. 599–620.

Bushnell, T.M. 1942. Some Aspects of the Soil Catena Concept. *Soil Science Society of America Proceedings.* Vol. 7, 466–76.

Butler, B.E. 1980. *Soil Classification for Soil Survey.* Oxford: Clarendon Press.

______. 1982. A New System for Soil Studies. *Journal of Soil Science.* Vol. 33, pp. 581–95.

Campbell, J.B. 1977. Variation of Selected Properties Across a Soil Boundary. *Soil Science Society of America Journal.* Vol. 41, pp. 578–82.

———. 1978. Spatial Variation of Sand Content and pH Within Single Contiguous Delineations of Two Soil Mapping Units. *Soil Science Society of America Journal.* Vol. 42, pp. 460–64.

———. 1979. Spatial Variability of Soils. *Annals of the Association of American Geographers.* Vol. 69, pp. 544–69.

———. 1981. Spatial Correlation Effects Upon Accuracy of Supervised Classification of Land Cover. *Photogrammetric Engineering and Remote Sensing.* Vol. 47, pp. 355–63.

———. 1983. *Mapping the Land: Aerial Imagery for Land Use Information.* Resource Publications in Geography. Washington, D.C.: Association of American Geographers.

Campbell, J.B., and W.J. Edmonds. 1984. The Missing Geographic Dimension of *Soil Taxonomy. Annals of the Association of American Geographers.* Vol. 74, pp. 83–97

Carver, Jonathan. 1802. *Three Years Travels Throughout Interior Parts of North America.* Charlestown.

Chamberlin, T.C. 1883. *Geology of Wisconsin.* Madison: Commissioners of Public Printing. 4 Vols.

Chorley, R.J., and P.F. Dale. 1972. Cartographic Problems in Stream Channel Delineation. *Cartography.* Vol. 7, pp. 150–62.

Chorley, R. J.; R. E. Beckinsale; and A.J. Dunn. 1973. *The History of the Study of Landforms or the Development of Geomorphology.* Volume II. *The Life and Work of William Morris Davis.* London: Methuen.

Chow, V.T. 1964 *Handbook of Applied Hydrology.* New York: McGraw-Hill.

Christian, C.S. 1959. The Eco-Complex in Its Importance for Agricultural Assessment. *Monographiae Biologicae* Vol. 8, pp. 587–605.

Christian, C.S., and G.A. Stewart. 1952. *A Survey of the Katherine-Darwin Region, 1946.* C.S.I.R.O. Land Research Series, No. 1. Melbourne, Australia.

Ciolkosz, E.J., S. Holzhey, B. Hajek, R. Rust, and L. Daugherty. 1982. Soil Characterization Data Studies. *Soil Survey Horizons.* Vol. 24, No. 4, pp. 11–13.

Clayden, B. 1971. *Soils of the Exeter District.* Memoirs of the Soil Survey of Great Britain, England, and Wales. Harpenden, Herts., England,

Cliff, A.D., and J.K. Ord. 1973. *Spatial Autocorrelation.* London: Pion.

Cline, M.G. 1944. Principles of Soil Sampling. *Soil Science.* Vol. 58, pp. 275–88.

———. 1949. Basic Principles of Soil Classification. *Soil Science.* Vol. 67, pp. 81–91.

———. 1963. Logic of the New System of Soil Classification. *Soil Science.* Vol. 96, pp. 17–22.

———. 1972. Thoughts About Appraising the Utility of Soil Maps. In *Soil Resource Inventories* (R.W. Arnold, ed.) Proceedings of a Workshop Organized by the Soil Resource Inventory Study Group at Cornell University, U.S.A. Agronomy Mimeo 77–23. Ithaca, New York: Department of Agronomy, Cornell University. Pp. 251–73.

Collins, J.B. 1972. Soil Resources Inventory for the Small Farmer. In *Soil Resource Inventories* (R.W. Arnold, ed.) Proceedings of a Workshop Organized by the Soil Resource Inventory Study Group at Cornell University, U.S.A. Agronomy Mimeo 77–23. Ithaca, New York: Department of Agronomy, Cornell University. Pp. 205–15.

Conacher, A.J., and J.B. Dalrymple. 1977. The Nine Unit Landscape Model: An Approach to Pedogeomorphic Research. *Geoderma.* Vol. 18, pp. 1–154.

Cooper, D.T. 1982. *Using a Geographic Data Base to Define Soil Boundaries.* Unpublished M.S. thesis. Columbia: University of Missouri.

Corrin, S. 1953. Dimensional Analysis and Similarity. In *Field Models in Geophysics* (Robert Long, ed.). Washington, D.C.: USGPO.

Crocker, R.L. 1952. Soil Genesis and the Pedogenic Factors. *The Quarterly Review of Biology.* Vol. 27, pp. 139–68.

Cruickshank, J.G. 1972. *Soil Geography.* Newton Abbot, England: David and Charles.

Davies, B.E., and S.A Gamm. 1970. Trend Surface Analysis Applied to Soil Reaction Values From Kent, England. *Geoderma.* Vol. 3, pp. 223–31.

Davies, P.C.W. 1977. *Space and Time in the Modern Universe.* Cambridge: Cambridge University Press.

Davis, John C. 1973. *Statistics and Data Analysis in Geology.* New York: John Wiley and Sons.

Delong, Alton J. 1981. Phenomological Space-Time: Toward an Experimental Relativity. *Science.* Vol. 213, pp. 681–83.

Dent, David, and Anthony Young. 1981. *Soil Survey and Land Evaluation.* Boston: Allen and Unwin.

Dokuchaev, V.V. 1883. *Russian Chernozem. (Selected Works of V.V. Dokuchaev:* Moscow, 1948). Translation by N. Kaner, Israel Program for Scientific Translation. (NTIS: TT 66–51133). Springfield, Va.: U.S. Department of Commerce.

______. 1886. *Evaluation of Soils in the Nizhniy Novogorod Government.* Independent Economics Society.

Doolittle, James A. 1982. Characterizing Soil Map Units With Ground-Penetrating Radar. *Soil Survey Horizons.* Vol. 23, No. 4, pp. 3–10.

Doolittle, James; Cornelius J. Heidt; Stuart L. Larson; Thomas P. Ryterske; Michael G. Ulmer; and Paul E. Welman. 1981. *Soil Survey of Grand Forks County, N.D.* Washington, D.C.: Soil Conservation Service, U.S.D.A.

Douglas, D.H., and Peucker, T.K. 1973. Algorithms for the Reduction of the Numbers of Points Required to Represent a Digitized Line or its Caricature. *The Canadian Cartographer.* Vol. 10, pp. 112–22.

Duchaufour, P. 1970. *Précis de Pédologie.* Paris: Masson.

Edmonds, W.J. 1983. *Grouping of Soil Profiles in Three Mapping Units by Conventional and Numerical Classifications.* Ph.D. dissertation, Virginia Polytechnic Institute, Blacksburg.

Engel, Robert J. 1977. *Soil Survey of Washtenaw County, Michigan.* Washington D.C.: Soil Conservation Service, U.S.D.A.

Erickson, A.J., and Lemoyne Wilson. 1968. Soil Survey of Davis-Weber Area, Utah. Washington, D.C.: Soil Conservation Service, U.S.D.A.

Faegri, K.; J. Iverson; and H.T. Waterbolk. 1964. *Textbook of Pollen Analysis.* Copenhagen: Munksgaard.

Fahey, L. 1954. *Generalization as Applied to Cartography.* Unpublished Master's thesis, Ohio State University.

Fischer, Roland. 1971. A Cartography of the Ecstatic and Meditative States. *Science.* Vol. 174, pp. 897–904.

Fosberg, Maynard. A. 1963. *Genesis of Some Soils Associated With Low and Big Sagebrush Complexes in the Brown, Chestnut, and Chernozem-Prairie Zones in South-Central and Southwestern Idaho.* Unpublished Ph.D. thesis, University of Wisconsin.

Fox, Robert E., and Gerhardt B. Lee. 1979. *Soil Survey of Dodge County, Wisconsin, U.S.A.* Washington, D.C.: Soil Conservation Service, U.S.D.A.

Fridland, V.M. 1974. Structure of the Soil Mantle. *Geoderma.* Vol. 12, pp. 35–41.

______. 1976a. *Pattern of the Soil Cover.* Moscow: Geographical Institute of the Academy of Sciences of the USSR, 1972. (Israel Program for Scientific Translation, 1976)

______. (editor) 1976b. *Soil Combinations and Their Genesis.* Translated from the Russian. Agricultural Research Service, U.S.D.A., and the National Science Foundation. New Delhi: Amerind Publishing Company.

______. 1980. Classification of the Structure of the Soil Mantle and Land Typification. *Soviet Soil Science.* 1980, pp. 642–54.

Forman, Richard T.T., and Michel Godron. 1981. Patches and Structure Components for a Landscape Ecology. *Bioscience.* Vol. 31, pp. 733–40.

Gaikawad, S.T., and F.D. Hole. 1961. Characteristics and Genesis of a Podzol Soil in Florence County, Wisconsin. *Transactions of the Wisconsin Academy of Science, Arts, and Letters.* Vol. 50, pp. 183–90.

Gennadiyev, A.N., and M.I. Gerasimov. 1980. Some Present Soil Classification Trends in the United States. *Soviet Soil Science.* Vol. 12, pp. 559–566.

Gerasimov, I.P. 1971. The Great Russian Scientist V.V. Dokuchaev on the 125th Anniversary of His Birth. *Soviet Soil Science.* Vol. 8, pp. 3–7.

Gerlach, A.C. (ed.) 1970. *The National Atlas of the United States.* Washington, D.C.: U.S. Geological Survey.

Gersmehl, P.J. 1977. Soil Taxonomy and Mapping. *Annals of the Association of American Geographers.* Vol. 67, pp. 419–28.

______. 1981. Maps in Landscape Interpretation. *Cartographia.* Vol. 18, No. 2, pp. 79–115.
Gersmehl, P.J., and D.E. Napton. 1982. Interpretation of Resource Data: Problems of Scale and Spatial Transferability. *Papers from the Annual Conference of the Urban and Regional Information Systems Association.* Pp. 471–82.
Getis, Arthur, and Barry Boots. 1978. *Models of Spatial Processes: An Approach to the Study of Point, Line, and Area Patterns.* London: Cambridge University Press.
Gile, Leland H. 1975. Causes of Soil Boundaries in an Arid Region: I. Age and Parent Materials. *Soil Science Society of America Proceedings.* Vol. 39, pp. 316–23. II. Dissection, Moisture, and Faunal Activity. Pp. 324–30.
Gile, Leland H., and Robert B. Grossman. 1979. *The Desert Soils Project Monograph.* Washington, D.C.: Soil Conservation Service, U.S.D.A.
Gile, L.H., and J.W. Hawley. 1972. The Prediction of Soil Occurrence in Certain Desert Regions of the Southwestern United States. *Soil Science of Society of America Proceedings.* Vol. 36, pp. 119–24.
Glinka, K.D. 1914. *Die Typen der Bodenbildung, Ihre Klassifikation und geographische Verbreitung.* Berlin.
______. 1927. *The Great Soil Groups of the World and Their Development.* (Translation by C.F. Marbut.) Ann Arbor, Michigan: Edwards.
Glocker, C.L., and R.A. Patzer. 1978. *Soil Survey of Dane County, Wisconsin.* Washington, D.C.: U.S.D.A. Soil Conservation Service.
Gray, D.S., and B.H. Hendrickson. 1928. *Soil Survey of Cherokee County, Iowa.* Washington, D.C.: U.S.D.A. Bureau of Chemistry and Soils.
Griffiths, T.M. 1971. *Scale Problems in Geographic Research.* Department of Geography, University of Denver. (U.S. Army ETL-CR-71-16)
Guptill, S.C. 1978. An "Optimal" Filter for Maps Showing Nominal Data. *Journal of Research, U.S. Geological Survey.* Vol. 6, pp. 161–67.
Habermann, G.M., and F.D. Hole. 1980. Soilscape Analysis in Terms of Pedogeomorphic Fabric: An Exploratory Study. *Soil Science Society of America Journal.* Vol. 44, pp. 336–40.
Hajek, B.F. 1972. Evaluation of Map Unit Composition by the Random Transect Method. In *Soil Resource Inventories* (R.W. Arnold, ed.) Proceedings of a Workshop Organized by the Soil Resource Inventory Study Group at Cornell University. Agronomy Mimeo. 77–23. Ithaca: Department of Agronomy, Cornell University. Pp. 71–87.
Hakanson, L. 1978. The Length of Closed Geomorphic Lines. *Mathematical Geology.* Vol. 10, pp. 141–67.
Hammond, E.H. 1962. Landform Geography and Landform Description. *California Geographer.* Vol. 3, pp. 71–72.
Harden, Jennifer W. 1982. A Quantitative Index of Soil Development From Field Descriptions: Examples From a Chronosequence in Central California. *Geoderma.* Vol. 28, pp. 1–28.
Harris, J.A. 1915. On a Criterion of Substratum Homogeneity (or Heterogeneity) in Field Experiments. *American Naturalist.* Vol. 49, pp. 430–54.
Harris, Stuart A. 1968. Landscape Analysis. In *The Encyclopedia of Geomorphology* (Rhodes W. Fairbridge, ed). Reinhold, New York. Pp. 626–29.
Hart, John Fraser. 1982. The Highest Form of the Geographer's Art. *Annals of the Association of American Geographers.* Vol. 72, pp. 1–29.
Hartshorne, Richard. 1939. The Nature of Geography: A Critical Survey of Current Thought in the Light of the Past. *Annals of the Association of American Geographers.* Vol. 29, pp. 173–645.
Hasse, G. 1968. Pedon und Pedotop—Bemerkungen zu Grundfragen der regionalen Bodengeographie. In *Landschadftsforschung. Beittrage zur Theorie und Anwendung* (H. Barthel, ed.). Leipzig: Hermann Haack, pp. 57–76.
______. 1973. Zur Ausgliederung von Raumeinheiten der chorischen under der regionischen Dimensiondergestellt an Biespielen der Bodengeographie. *Petermanns Geographische Mitteilungen.* Vol. 117, No. 2, pp. 81–90.
Hathaway, L.R.; T.C. Waugh; O.K. Galle; and H.P. Dickey. 1979. *Chemical Quality of Irrigation Waters in Northwestern Kansas.* Chemical Quality Series 8. Kansas Geological Survey, Lawrence.

Hilgard, E.G. 1880. Report on Cotton Production in the U.S.: Also Embracing Agricultural and Physiographic Description of the Several Cotton States and California. *Tenth Census of the United States.* Washington, D.C.: U.S. Census Office.

———. 1906. *Soils: Their Formation, Properties, Composition, and Relation to Climate and Plant Growth.* New York: Macmillan.

Hills, R.C., and S.G. Reynolds. 1969. Illustrations of Soil Moisture Variability in Selected Areas and Plots of Different Sizes. *Journal of Hydrology.* Vol. 8, pp. 27–47.

Hinckley, K.C. 1978. *Soil Survey of DeKalb county, Illinois.* Washington, D.C.: Soil Conservation Service, U.S.D.A.

Hodges, R.L.; H.L. Mathews; D.F. Amos; J.P. Sutton; and W.J. Edmonds. 1978. *Soil Survey of Chesterfield County, Virginia.* Washington, D.C.: Soil Conservation Service, U.S.D.A.

Hole, F.D. 1953. Suggested Terminology for Describing Soils as Three-Dimensional Bodies. *Soil Science Society of America Proceedings.* Vol. 17, pp. 131–35.

———. 1961. A Classification of Pedoturbations and Some Other Processes and Factors of Soil Formation in Relation to Isotropism and Anisotropism. *Soil Science.* Vol. 91, pp. 375–77.

———. 1975. Some Relationships Between Forest Vegetation and Podzol Horizons in Soils of Menominee Tribal Lands, Wisconsin, U.S.A. *Soviet Soil Science.* Vol. 7, No. 6, pp. 714–723.

———. 1976. *Soils of Wisconsin.* Madison: University of Wisconsin Press.

———. 1978. An Approach to Landscape Analysis with Emphasis on Soils. *Geoderma.* Vol. 21, pp. 1–23.

———. 1981. Effects of Animals on Soil. *Geoderma.* Vol. 25, pp. 75–112.

———. 1983. A Soilscape Analysis of Some Geomorphic Surfaces in the State of Iowa, U.S.A. *Abstracts, 17th Annual Meeting, Geological Society of America, North-Central Section.* Vol. 15, p. 212.

Howell, J.V. 1957. *Glossary of Geology and Related Sciences.* American Geological Institute.

Huggett, R.J. 1975. Soil Landscape Systems: A Model of Soil Genesis. *Geoderma.* Vol. 13, pp. 1–22.

Hutchins, Thomas. 1778. *A Topographical Description of Virginia, Pennsylvania, Maryland, and North Carolina.* London.

Isachenko, A.G. 1973. Landscape Science. In *Great Soviet Encyclopedia* (A.M. Prokhorov, ed.). New York: Macmillan Educational Corporation. Vol. 14, pp. 195–96.

Jenks, G.F. 1981. Lines, Computers, and Human Frailties. *Annals, Association of American Geographers.* Vol. 71, pp. 1–10.

Jenny, Hans. 1941. *Factors of Soil Formation.* New York: McGraw-Hill.

———. 1961. *E.W. Hilgard and the Birth of Modern Soil Science.* Pisa, Italy: Agrochimica; and Berkeley, California: Farallon.

———. 1965. Tessera and Pedon. *Soil Survey Horizons.* Vol. 6, pp. 8–9.

———. 1980. *The Soil Resource: Origins and Behavior.* Berlin: Springer-Verlag.

Jensen, D.R., and J. Pesek. 1962. Inefficiency of Fertilizer Use Resulting From Nonuniform Spatial Distribution: I.Theory. II.Yield Losses Under Selected Distribution Patterns. III. Fractional Segregation in Fertilizer Materials. *Soil Science Society of America Proceedings,* Vol. 26, pp. 170–82.

Johnson, W.D. 1963. The Pedon and the Polypedon. *Soil Science Society of America Proceedings.* Vol. 27, pp. 212–15.

Jones, T.A. 1959. Soil Classification—A Destructive Criticism. *Journal of Soil Science.* Vol. 10, pp. 196–200.

Kalm, Per. 1770. *Travels in North America.* London.

Kellogg, C.E., and A.C. Orvedal. 1969. Potentially Arable Soils of the World and Critical Measures for Their Use. *Advances in Agronomy.* Vol. 21, pp. 109–70.

King, G.M. 1982. Relation of Soil Water Movement and Sulphide Concentration to *Spartina alterniflora* Production in a Georgia Salt Marsh. *Science.* Vol. 218, pp. 61–63.

Kir'Yanov, G.F. 1965. Some Problems of Gnosiology in the Works of V.V. Dokuchaev. *Soviet Soil Science.* Vol. 1965, pp. 1.

Klingelhoets, A.J. 1962. *Soil Survey of Iowa County, Wisconsin.* Washington, D.C.: Soil Conservation Service, U.S.D.A.

Knox, E.G. 1965. Soil Individuals and Soil Classification. *Soil Science Society of America Proceedings.* Vol. 29, pp. 79–84.

Krusekopf, H.H. 1943. *Life and Work of C.F. Marbut, Soil Scientist.* Memorial Volume. Columbia, Missouri: Soil Science Society of America.

Laker, M.C. 1972. Selection of a Method for Determination of Map Intensities of Soil Maps. In *Soil Resource Inventories* (R.W. Arnold, ed.). Proceedings of a Workshop Organized by the Soil Resource Inventory Study Group at Cornell University. Agronomy Mimeo 77–23. Ithaca: Department of Agronomy, Cornell University, pp. 125–39.

Larson, W.E.; F.J. Pierce; and R.H. Dowdy. 1983. The Threat of Soil Erosion to Long-Term Crop Production. *Science.* Vol. 219, pp. 458–65.

Legros, J.P. 1978. Recherche et contrôle numérique de la précision en cartographie pédologique. I. Précision dans la dé limitation des sols. *Annales Agronomique.* Vol. 29, pp. 499–519. II. Précision dans la caracterisation des unités de sols. Pp. 583–601.

Lepsch, I.F.; S.W. Buol; and R.B. Daniels. 1977. Soil Landscape Relationships in the Occidental Plateau of Sao Paulo State, Brazil: Geomorphic Surfaces and Soil Mapping Units. *Soil Science Society of America Journal.* Vol. 41, pp. 104–109.

Link, Ernest G.; Steven L. Elmer; and Sidney A. Vanderveen. 1978. *Soil Survey of Door County, Wisconsin.* Washington, D.C.: Soil Conservation Service, U.S.D.A.

Lyford, W.H. 1974. Narrow Soils and Intricate Soil Patterns in Southern New England. *Geoderma.* Vol. 11, pp. 195–208.

Mace, T.H. 1980. *Microdensiotometric Analysis of Aerial Photographic Imagery for Detailed Soil Maps.* Unpublished Ph.D. thesis, University of Wisconsin, Madison.

Mandelbrot, Benoit. B. 1983. *The Fractal Geometry of Nature.* New York: W.H. Freeman.

Marbut, C.F. 1935. *Soils of the United States.* Part III. In *Atlas of American Agriculture.* Washington, D.C.: U.S.D.A. Pp. 1–98.

______. 1951 (1928). *Soils: Their Genesis and Classification.* Madison, Wisconsin: Soil Science Society of America.

McBratney, A.; R. Webster; R.G. McLaren; and R.B. Spiers. 1982. Regional Variation of Extractable Copper and Cobalt in the Topsoil of South-East Scotland. *Agronomie.* Vol. 2, pp. 969–82.

Milfred, C.J., and R.W. Keifer. 1976. Analysis of Soil Variability with Repetitive Aerial Photography. *Soil Science Society of America Journal.* Vol. 40, pp. 553–57.

Milfred, C.J.; G.W. Olson; and F.D. Hole. 1967. *Soil Resources and Forest Ecology of Menominee County, Wisconsin.* Madison: University of Wisconsin Geological and Natural History Survey Bulletin 85.

Milne, G. 1935. Composite Units for Mapping Complex Soil Associations. *Third International Congress of Soil Science.* Vol. 1. Oxford, England.

Mitchell, M.J. 1980. *Soil Survey of Winnebago County, Wisconsin, U.S.A.* Washington, D.C.: Soil Conservation Service, U.S.D.A.

Mockma, D.L.; E.P. Whiteside; and E.F. Schneider. 1972. *Soils of Oceana County, Michigan.* (in two volumes). Dept. of Crop and Soil Science, Michigan State University, East Lansing.

Mockma, D.L., and E.P. Whiteside. 1973. *Soils of Newaygo County, Michigan.* (in two volumes). Dept. of Crop and Soil Science, Michigan State University, East Lansing.

Moholy-Nagy, Sibyl. 1966. The Four Environments of Man. *Landscape.* Vol. 16, pp. 3–9.

Monkhouse, F.J. and H.R. Wilkinson. 1971. *Maps and Diagrams.* London: Methuen.

Moran, Emilio F. 1981. *Developing the Amazon.* Bloomington: Indiana University Press.

Mueller, Jerry E. 1979. Problems in the Definition and Measurement of Stream Length. *Professional Geographer.* Vol. 31, pp. 306–11.

Nichols, J.D. 1975. Characteristics of Computerized Soil Maps. *Soil Science Society of America Proceedings.* Vol. 39, pp. 927–32.

Nisbet, Jerry J. 1979. The Brain, The Beauty of Science and Learning. *Proceedings, Indiana Academy of Sciences.* Vol. 88, pp. 52–57.

Nwadialo, Bernard-Shaw E. 1978. *Statistical Analysis of Two Soil Properties Along Two Transects on Some Forested Soil Bodies in Wisconsin.* Unpublished Master's thesis, University of Wisconsin, Madison.

Nygard, I.J., and F.D. Hole. 1949. Soil Classification and Soil Maps: Units of Mapping. *Soil Science.* Vol. 67, pp. 163–68.

Oliver, C.R.; A.P. Bell; T.E. Barnes; S. Meyers; and A.T. Meyers. 1946. *Soil Survey of Fulton County, Indiana.* U.S.D.A. Bureau of Plants, Industry, Soils, and Agricultural Engineering.

Olson, Jerry. 1980. Forward: Hans Jenny and Fertile Soil. In *The Soil Resource: Origin and Behavior* (Hans Jenny). New York: Springer-Verlag. Pp. vii-xiii.

Olson, J.R.; L.P. Wilding; and G.F. Hall. 1980. Soil Taxonomy and Regional Land Use Interpretations for Ohio. *Soil Science Society of America Journal.* Vol. 44, pp. 1069–74.

Orvedal, A.C., and M.J. Edwards. 1941. General Principles of Technical Grouping of Soils. *Soil Science Society of America Proceedings.* Vol. 6, pp. 386–91.

Orvedal, A.C.; M. Baldwin; and A.J. Vessel. 1949. Soil Classification and Soil Maps: Compiled Maps. *Soil Science.* Vol. 67, pp. 177–81.

Oschwald, W.R. 1965. *The Effect of Size and Shape of Soil Mapping Unit in Determining Soil Use Potential.* Ph.D. dissertation, Iowa State University. Ann Arbor, Michigan: University Microfilms 66–3003.

Pastor, John; John D. Aber; and Charles A. McClaughtery. 1982. Geology, Soils, and Vegetation of Blackhawk Island, Wisconsin. *The American Midland Naturalist.* Vol. 108, pp. 266–77.

Pavlick, Hannah H. 1974. *The Attributes of Soilscapes—A Preliminary Analysis.* Unpublished Master's thesis, University of Wisconsin, Madison.

Pavlick, H., and F.D. Hole. 1977. Soilscape Analysis of Slightly Contrasting Terrains in Southeastern Wisconsin. *Soil Science Society of America Proceedings.* Vol. 41, pp. 407–13.

Peck, T.R., and S.W. Melsted. 1967. *Field Sampling for Soil Testing.* Soil Science Society of America Special Publication 2 (Part 1). Madison, Wisconsin. Pp. 25–35.

Peterson, Frederick F. 1981. *Landforms of the Basin and Range Province Defined for Soil Survey.* Technical Bulletin 28, Nevada Agricultural Experiment Station, University of Nevada, Reno.

Piaget, J., and B. Inhelder. 1967. *The Child's Conception of Space.* Translated from the French by F.J. Langdon and J.J. Lunzer. New York: Norton.

Pitty, A.F. 1979. *Geography and Soil Properties.* Cambridge: Methuen.

Poore, J.K., and J.H. Huddleston. 1983. Shape Analysis of Soil Mapping Units. *Agronomy Abstracts.* Madison, Wisconsin: American Society of Agronomy. Pp. 188–89.

Prokhorov, A.M. 1973. *Great Soviet Encyclopedia.* New York: Macmillan Educational Corporation.

Putnam, D.F. 1951. Soils and Their Geographic Significance. Chapter 10 in *Geography in the Twentieth Century* (G. Taylor, ed). London: Methuen. Pp. 221–247.

Reed, J.F., and J.A. Rigney. 1947. Sampling From Fields of Uniform and Non-Uniform Appearance and Soil Types. *Journal of the American Society of Agronomy.* Vol. 39, pp. 26–40.

Riecken, F.F. 1963. Some Aspects of Soil Classification in Farming. *Soil Science.* Vol. 96, pp. 49–61.

Rieder, Neil E.; Victor L. Riemenschneider; and Paul W. Reese. 1973. *Soil Survey of Ashtabula County, Ohio, U.S.A.* Washington, D.C.: Soil Conservation Service, U.S.D.A.

Robinove, C.J. 1981. The Logic of Multispectral Classification and Mapping of Land. *Remote Sensing of Environment.* Vol. 11, pp. 231–44.

Robinson, A.; R. Sale; and J. Morrison. 1978. *Elements of Cartography.* New York: John Wiley and Sons.

Robinson, Arthur H., and Barbara B. Petchenik. 1976. *The Nature of Maps: Essays Toward Understanding Maps and Mapping.* Chicago: University of Chicago Press.

Rode, A.A. 1962. *Soil Science.* Translated from the Russian by the Israel Program for Scientific Translation. Jerusalem. National Science Foundation. Washington, D.C.

Rogers, Oliver C.; A.P. Bell; T.E. Barnes; S. Meyers; and A.T. Wianko. 1946. *Soil Survey of Fulton County, Indiana.* Washington, D.C.: U.S.D.A. Bureau of Chemistry and Soils.

Ruhe, R.V. 1969. *Quaternary Landscapes in Iowa.* Ames: Iowa State University Press.

———. 1974. *Geomorphology.* Boston: Houghton-Mifflin.

Ruhe, R.V.; Daniels and Cady: 1967. *Landscape Evolution and Soil Formation in Southwestern Iowa.* U.S.D.A. Technical Bulletin 1349. Washington, D.C.

Sallee, K.E. 1972. *Soil Survey of Lane County Kansas.* Washington, D.C.: Soil Conservation Service, U.S.D.A.

Sanchez, Pedro A.; Walter Couto; and Stanley W. Buol. 1982. The Fertility-Capability Soil Classification System: Interpretation, Applicability, and Modification. *Geoderma.* Vol. 27, pp. 283–309.

Sanders, F.W.; H.R. Sinclair; and H.M. Galloway. 1979. History of the Miami Series in Indiana. *Proceedings, Indiana Academy of Science.* Vol. 88, pp. 405–10.

Schelling, J. 1970. Soil Genesis, Soil Classification, and Soil Survey. *Geoderma.* Vol. 4, pp. 165–93.

Schlichting, E. 1970. Bodensystematik und Bodensoziologie. *Zeitschrift für Pflanzenernährung und Bodenkunde.* Vol. 127, No. 1, pp. 1–9.

Sibirtsev, N.M. (S.S. Sobolev, editor). 1914. *Selected Works. Vol. 1: Soil Science.* (Translation 1966. NTIS TT 65–50126). Reprinted Moscow, 1951. Vol. 2: Soil Science and Drought Control. (Translation 1966. NTIS: 66–51136; 392 pp.)

Simonson, R.W. 1959. Outline of a Generalized Theory of Soil Genesis. *Soil Science Society of America Proceedings.* Vol. 23, pp. 152–56.

______. 1968. Concept of Soil. *Advances in Agronomy.* Vol. 20, pp. 1–47.

______. 1971. Soil Association Maps and Proposed Nomenclature. *Soil Science Society of America Proceedings.* Vol. 35, pp. 959–65.

______. 1978. A Multiple-Process Model of Soil Genesis. In *Quaternary Soils* (W.C. Mahaney, ed.). Norwich, England: Geo Abstracts, University of East Anglia., pp. 1–25.

Simonson, R.W.; F.F. Riecken; and G.D. Smith. 1952. *Understanding Iowa Soils.* Dubuque, Iowa: W.C. Brown.

Sklar, Lawrence. 1974. *Space, Time, and Spacetime.* Berkeley: University of California Press.

Slota, Robert W., and Glenn D. Garvey. 1961. *Soil Survey of Crawford County, Wisconsin.* Washington, D.C.: Soil Conservation Service, U.S.D.A.

Smeck, Niel E., and E.C.A. Runge. 1971. Phosphorus Availability and Redistribution in Relation to Profile Development in an Illinois Landscape Segment. *Soil Science Society of America Proceedings.* Vol. 35, pp. 953–59.

Smith, Guy D., and Michael Leamy. 1982. Conversations in Taxonomy. *Soil Survey Horizons.* Vol. 2, No. 3, pp. 4–17.

Soil Conservation Service. 1967. *Distribution of Principal Kinds of Soils: Orders, Suborders, and Great Groups.* Sheet 85 in *National Atlas of the United States* (A. Gerlach, ed.). Washington: U.S. Geological Survey.

Soil Conservation Service, U.S.D.A. 1979. *General Soil Map: Virginia.*

Soil Survey Staff. 1951. *Soil Survey Manual.* Agriculture Handbook No. 18. Washington, D.C.: U.S.D.A.

______. 1960. *Soil Classification: A Comprehensive System. 7th Approximation.* Washington, D.C.: Soil Conservation Service, U.S.D.A.

______. 1975a. *Soil Taxonomy: A Basic System for Making and Interpreting Soil Surveys.* Agriculture Handbook No. 436. Washington, D.C.: U.S.D.A.

______. 1975b. *Revised Soil Survey Manual* (draft). Washington, D.C.: Soil Conservation Service, U.S.D.A.

______. 1980. *Classification of Soil Series in the United States.* Washington, D.C.: Soil Conservation Service, U.S.D.A.

Soil Science Society of America. 1979. *Glossary of Soil Science Terms.* Madison, Wisconsin.

Spencer, J.E. 1966. *Shifting Cultivation in Southeastern Asia.* Berkeley: University of California Publications in Geography. Vol. 19.

Spencer, J.E., and G.A. Hale. 1961. The Origin, Nature, and Distribution of Agricultural Terracing. *Pacific Viewpoint.* Vol. 2, pp. 1–41.

Steingraeber, J.A., and C.A. Reynolds. 1971 *Soil Survey of Milwaukee and Waukesha Counties, Wisconsin.* Washington, D.C.: Soil Conservation Service, U.S.D.A.

Strickland, William. 1801 (1971). *Journal of a Tour in the United States of America and Observations on the Agriculture of the United States of America.* New York.

Swenson, John L.; Delyle Becksrand; Dwaine T. Erikson; Calvin McKinley; Jungi J. Shiozaki; and Ronald Tew. 1982. *Soil Survey of Sanpete Valley Area, Utah: Parts of Sanpete and Utah Counties.* Washington, D.C.: Soil Conservation Service, U.S.D.A.

Tanner, C.B., and J. Bouma. 1975. *Evapotranspiration as a Means of Domestic Liquid Waste Disposal in Wisconsin.* Madison: Geological and Natural History Survey, University of Wisconsin.

Terzaghi, K., and R.B. Peck. 1948. *Soil Mechanics in Engineering Practice.* New York: John Wiley and Sons.

Thomas, M.F. 1969. Geomorphology and Land Classification in Tropical Africa. In *Environment and Land Use in Africa* (M.F. Thomas, and G.W. Whittington, eds.). London: Methuen.

Thomasson, A.J. 1971. *Soils of the Melton Mowbray District.* Memoirs of the Soil Survey of Great Britain, England, and Wales. Harpenden, Herts., England.

Troeh, F.R. 1964. Landform Parameters Correlated to Soil Drainage. *Soil Science Society of America Proceedings.* Vol. 28, pp 808–12.

Troeh, Frederick R. 1975. Measuring Soil Creep. *Soil Science Society of America Proceedings.* Vol. 39, pp. 707–9.

Tuan, Yi-Fu. 1974. *Topophilia: A Study of Environmental Perceptions, Attitudes, and Values.* Englewood Cliffs: Prentice-Hall.

______. 1977. *Space and Place: The Perspective of Experience.* Minneapolis: Univeristy of Minnesota Press.

Tyurin, L.V.; I.P. Gerasimov; E.N. Ivanova; and V.A. Nosin (editors). *Soil Survey: A Guide to Field Investigations and Mapping of Soils.* Academy of Sciences of the U.S.S.R., V.V. Dokuchaev Soil Institute. Translated from the Russian, 1965. Jerusalem: Israel Program for Scientific Translation. (U.S. Department of Commerce, Springfield, Va.)

Ulrich, H.P.; T.E. Barnes; S. Meyers; O.C. Rogers; and A.T. Wiancho. 1947 *Soil Survey, Bartholomew County, Indiana.* Washington: U.S.D.A. Bureau of Plant Industry, Soils, and Agricultural Engineering.

U.S. Forest Service. 1976. *Land Systems Inventory Guide.* Missoula, Montana: U.S. Forest Service.

U.S. Salinity Laboratory Staff. 1954. *Diagnosis and Improvement of Saline and Alkali Soils.* Agriculture Handbook 60. Washington, D.C.: U.S.D.A.

Vance, James E. 1980. Ordinary Landscapes. A review of D.W. Meinig (ed.) *The Interpretation of Ordinary Landscapes: Geographical Essays.* New York: Oxford University Press *Landscape.* Vol. 24, No. 2, pp. 29–33.

Van Heesen, H.C. 1970. Presentation of the Seasonal Fluctuation of the Water Table on Soil Maps. *Geoderma.* Vol. 4, pp. 257–78.

Van der Voet, D. 1959. *Soil Survey, Rockingham County, New Hamphshire.* Soil Conservation Service, U.S.D.A. (Series 1954, No. 5).

Van Rooyen, D.J. 1972. *Soils of the Lake Wingra Basin.* Unpublished Master's thesis, University of Wisconsin, Madison. Pp. 85–147.

Van Wambeke, A. 1966. Soil Bodies and Soil Classifications. *Soils and Fertilizers.* Vol. 29, pp. 507–10.

Varnes, D.J. 1974. *The Logic of Geological Maps With Reference to Their Interpretation for Engineering Purposes.* Geological Survey Professional Paper 837, Washington, D.C.,

Veatch, J.O.; L.R. Schoenmann; Z.C. Foster; and F.R. Lesh. 1927. *Soil Survey of Kalkaska County Michigan.* U.S.D.A. Bureau of Chemistry and Soils.

Veatch, J.O.; L.R. Schoenmann; A.L. Gray; C.S. Simmons; and Z.C. Foster. 1927. *Soil Survey of Chippewa County, Michigan.* U.S.D.A. Bureau of Chemistry and Soils.

Viessman, W.; T.E. Harbaugh; and J.W. Knapp. 1972. *Introduction to Hydrology.* New York: Intext Educational Publishers.

Walker, P.H.; G.F. Hall; and R. Protz. 1968. Relation Between Landform Parameters and Soil Properties. *Soil Science Society of America Proceedings.* Vol. 32, pp. 101–4.

Walter, H., and H. Straha. 1970. *Arealkunde, Florischtiche-Historische Geobotanik.* Stutgart.

Webster, R. 1968. Fundamental Objections to the 7th Approximation. *Journal of Soil Science.* Vol. 19, pp. 354–66.

______. 1978. Optimally Partitioning Soil Transects. *Journal of Soil Science.* Vol. 29, pp. 388–402.

Webster, R., and P.H.T. Beckett. 1968. Quality and Usefulness of Soil Maps. *Nature.* Vol. 219, pp. 680–82.

Webster, R., and P.A. Burrough. 1972. Computer-Based Soil Mapping of Small Areas From Sample Data: I. Multivariate Classification and Ordination. *Journal of Soil Science.* Vol. 23, pp. 210–21. II. Classification Smoothing. Pp. 222–34.

Westin, Frederick F. 1981. Geography of Soil Test Results. *Soil Science Society of America Journal.* Vol. 40, pp. 890–95.

Wehde, M. 1982. Grid Cell Size in Relation to Errors in Maps and Inventories Produced by Computerized Map Processing. *Photogrammetric Engineering and Remote Sensing.* Vol. 48, pp. 1289–98.

Whitehead, A.N. 1925. *Science and the Modern World.* New York: Macmillan.

Whittig, L.D., and P. Janitzky. 1963. Mechanisms of Formation of Sodium Carbonate in Soils. I. Manifestations of Biological Conversions. *Journal of Soil Science.* Vol. 14, pp. 322–33.

Wildermuth, R., and L. Kraft. 1926. *Soil Survey of Kent County, Michigan.* U.S.D.A. Bureau of Chemistry and Soils.

Wilding, L.P., and L.P. Miller. 1979. Confidence Limits for Soil Survey. *Proceedings of the National Technical Work-Planning Conference of the Cooperative Soil Survey.* San Antonio, Texas, pp. 72–98.

Williams, B.J., and C.A. Ortiz-Solario. 1981. Middle American Folk Soil Taxonomy. *Annals of the Association of American Geographers.* Vol. 71, pp. 335–58.

Wilson, M.A.; L.A. Zelazny; and J.C. Baker. 1983. *An Investigation of Soils within the Tatum and Elioak Mapping Units in the Virginia Piedmont.* Blacksburg: Virginia Agricultural Experiment Station. Bulletin 83–1.

Worster, John R.; Elmer H. Harvey; and Laurence T. Hanson. 1972. *Soil Survey of Woodbury County, Iowa, U.S.A.* Washington, D.C.: U.S. Department of Agriculture, Soil Conservation Service.

Youden, W.J., and A. Mehlich. 1937. Selection of Efficient Methods for Soil Sampling. *Contributions From Boyce Thompson Institute.* Vol. 9, pp. 59–70.

Zakrzewska, Barbara. 1967. Trends and Methods in Land Form Geography. *Annals of the Association of American Geographers.* Vol. 57, pp. 129–65.

Zaslavsky, D., and A.S. Rogowski. 1969. Hydrologic and Morphologic Implications of Anisotropy and Infiltration in Soil Profile Development. *Soil Science Society of America Proceedings.* Vol. 33, pp. 594–99.

Index

About the Authors

FRANCIS D. HOLE, Emeritus Professor of Soil Science and Geography, University of Wisconsin, Madison, holds a B.A. degree from Earlham College, an M.A. Degree from Haverford College, and a Ph.D. Degree from the University of Wisconsin. Dr. Hole is a Fellow of the Soil Science Society of America, the American Society of Agronomy, the Geological Society of America, the American Association for the Advancement of Science, and the Soil Conservation Society of America. Dr. Hole has coauthored *Soil Genesis and Classification, Soil Science Simplified, Soils of Wisconsin, Soil Resources and Forest Ecology of Menominee County, Wisconsin,* as well as many county soil survey maps and reports. He has also published articles in various journals, including *Soil Science Society of America Journal, Soil Science,* and *Transactions of the Wisconsin Academy of Sciences, Arts and Letters.*

JAMES B. CAMPBELL, Associate Professor of Geography at Virginia Polytechnic Institute and State University, Blacksburg, holds a B.A. degree from Dartmouth College, and M.A. and Ph.D. degrees from the University of Kansas, Lawrence. He is a member of the Soil Science Society of America, the British Society of Soil Science, the International Association of Mathematical Geologists, the Association of American Geographers, and Sigma Xi. At Virginia Polytechnic Institute he teaches courses in Physical Geography, Cartography, and Remote Sensing. His work has been published in journals devoted to soil science, geography, geology, and remote sensing. He is the author of *Mapping the Land: Aerial Imagery for Land Use Information.*